Mead Making for Beginners

The Complete Guide to Crafting Your Mead at Home, from Basic Brewing to Advanced, with Essential Tips and Techniques. | BONUS: Beginner-Friendly Recipes

Michael York

Contents

YOUR FREE GIFTS!

In the spirit of deepening your journey into the craft of mead-making, I'm thrilled to offer not just one but two exclusive bonuses designed to enrich your brewing adventure. First, allow me to introduce the **Mead Making Tracker**, a meticulously crafted tool to accompany you every step of the way. This tracker ensures no detail, from selecting your water source to the precise timing of nutrient additions, escapes your notice. Its organized format lets you document each ingredient, step, and observation, elevating your mead-making process to a veritable art form.

But the journey doesn't end there. To further ignite your passion and expand your repertoire, I've also prepared a second bonus – a small book featuring **10 additional beginner-friendly recipes** that complement and extend beyond the techniques and recipes that I will share with you later in the book. Each recipe is designed to introduce new flavors and techniques, encouraging exploration and creativity within the bounds of traditional mead-making.

In addition to these two invaluable resources, I am excited to present a **third bonus**—a detailed guide on crafting mead in the time-honored tradition using **wild yeasts** and **all-natural ingredients**, allowing you to explore the roots of mead-making in its most authentic form.

SCAN THE QR CODE ABOVE TO DOWNLOAD YOUR FREE GIFTS!

INTRODUCTION

Carboys full of mead/Credit: Scott Wilson (www.Shutterstock.com)

Have you ever marveled at the rich, historical range of ancient beverages and wondered if you could, perhaps, enjoy being part of that tradition of tasty drinks? Or perhaps you've considered brewing your own beer or wine, but were daunted by the complex skills and physical space required?

Well, here´s an excellent, relevant and exciting option. Mead-making.

It offers a unique, less-explored avenue in home-made beverages. It combines the simplicity of blending ingredients with the richness of tradition—all without needing to set up a vineyard or brewery!

Mead, often revered as the ancestor of all fermented drinks, requires water, honey, and yeast. Unlike a home brewery, mead-making doesn't demand a vast space, or disrupt your neighbors with noise. If time is your concern, you'll be pleased to know that mead-making can be as hands-on—or laid-back—as you prefer. Once the fermentation process begins, the magic primarily unfolds on its own, with only occasional guidance from you.

If you're holding this book, it's likely you're curious about learning this ancient craft, transforming simple ingredients into something delightful to taste—and rich in history.

You're about to become part of a community that values sustainability, craftsmanship, and the joy of creating something special and your very own.

In this book, I'll guide you through the fascinating process of turning honey into mead. From selecting the right ingredients to understanding the fermentation process and bottling your first batch, this book is designed to integrate seamlessly into your life, whether mead-making becomes a cherished hobby or a burgeoning business.

So, whether you're a seasoned homebrewer looking to explore new horizons; a history enthusiast eager to recreate a drink from the past; or someone looking for a rewarding hobby that brings a sense of accomplishment, *Mead Making for Beginners* has something to offer you all.

Let's embark on this journey together, crafting delicious meads that celebrate nature's simplicity and flavor's complexity.

Book Structure Overview

This book has been thoughtfully designed to suit newcomers to the ancient art of mead-making, regardless of any prior experience. I will guide you towards becoming a mead-maker who is confident, versatile, and capable of producing both straightforward and complex varieties of mead. I will (hopefully) motivate and empower you through engaging and practical advice, encouraging a responsible and cautious approach to this rewarding craft. Here's a glimpse into the treasures you'll discover in this book:

- **Your Questions Answered:** Address the budding mead-maker's curiosity and concerns, covering the legality of home brewing, the anticipated time from brewing to tasting, and insights into scaling a hobby into a commercial venture.

- **Styles Unveiled:** Dive deep into the rich tapestry of mead styles, from traditional to experimental, categorizing them by sweetness levels, ingredients, and fermentation techniques. This section aims to familiarize readers with the versatility and range of mead.

- **Ingredients Deep Dive:** A comprehensive guide to the four pillars of mead-making: water, honey, yeast, and nutrients. Explore the nuances of selecting the right honey variety, the importance of water quality, choosing the optimal yeast strain, and the critical role of nutrients in fermentation health.

- **Equipment Essentials:** Start with the basics, detailing the equipment for your first batch, and gradually introduce more sophisticated tools and technology to enhance the mead-making process, as your skills flourish.

- **The Making Process:** A comprehensive, step-by-step walkthrough of mead production, from initial sanitation, preparing the must, pitching the yeast, through primary and secondary fermentation phases, to the final bottling, including strategies like Staggered Nutrient Addition and degassing.

- **Advanced Techniques:** Elevate your mead with techniques for stabilization, back sweetening, clarification, aging, oaking, adjusting acidity, carbonation, and more. Each method is broken down to help you refine the character and quality of your mead.

- **Crafting with Additives:** Unleash your creativity by incorporating fruits for melomels, apples for cysers, grapes for pyments, and spices or herbs for metheglins. Learn how to blend honey with malt for braggots, detailing the process for each variant.

- **Recipe Development:** Master the art of recipe creation, from conceptualizing the mead's profile to precise nutrient calculations for optimal fermentation. This section provides tools and techniques for aspiring recipe developers to achieve their desired outcomes.

- **Beginner-Friendly Recipes:** Kickstart your mead-making lifestyle with 15 meticulously crafted recipes designed for beginners. These recipes practically apply the book's teachings, guiding you through creating diverse and delicious meads.

What does mead taste like?

When people first encounter mead, there's often the expectation of sweetness, a logical assumption given its honey base. However, mead's taste landscape is surprisingly varied. It's not *just* about sweetness; Mead can be as dry as a desert, or richly sweet. It embraces diversity in forms like melomel—bursting with fruity notes; or metheglin—enriched with bold spices. The world of mead is a canvas for endless flavor-painting possibilities!

The unmistakable essence of fermented honey is at the core of mead's identity. The type of honey used doesn't just contribute a flavor; it defines mead's character. Imagine honey's essence, without sweetness—this captures a traditional mead's profile. It should remind you of floral bouquets, clean and refined, like a white wine—but with a more pronounced flavor.

The Common Query: Honey Wine or Mead?

Let's clear up a typical puzzle in the world of mead-making: Is there a fundamental difference between honey wine and mead? Well, the answer is as sweet and straightforward as the drink itself. Both terms refer to the same delightful beverage. It's a bit like deciding whether to call a beloved grandparent, "Grandma" or "Nana"—the choice is yours, and either way, it's all about affection and tradition.

Think of "mead" as the more popular term, especially when we're trying to distinguish it from its cousins—beer, and wine. Beer, born from grain, and wine, from fruits, stand in contrast to mead, which owes its existence to honey. But if you fancy calling it honey wine, you're equally on point. Historically, these terms have danced around each other, sometimes leading; sometimes following, but always in step with the essence of the drink.

Now, let's talk about a twist in the tale—the legal side of things in the U.S. The Alcohol & Tobacco Tax and Trade Bureau (TTB), the big boss in alcohol regulation, historically preferred "honey wine" over "mead" on labels. This was the norm until 2016. Imagine a world where "mead" wasn't even allowed on labels! Confusing, right? But here's where it gets better. The American Mead Makers Association stepped in, and now, both "mead" and "honey wine" happily co-exist on labels. So, whether you're a "mead" person or a "honey wine" enthusiast, the law's got your back.

And here's a very heartwarming bit. Whether you say mead or honey wine, you're toasting a beverage with an extraordinary history and a culinary footprint that spans the globe. It's a drink that's been part of human celebrations, sorrows, and stories for literally ages. So, the next time you raise a glass of this honey-infused delight, remember you're not just sipping on a drink—but savoring a piece of human history.

Unveiling the Mystique: Wine vs. Mead

Diving into the Ingredients: The Heart of the Matter

Let's start by unraveling a common misconception: Mead is not just a type of wine known as "honey wine"; it's in a unique category. The main difference? It's all in the ingredients. Mead is born from the fermentation of honey, while wine is the elegant result of fermenting grape juice. It's like comparing apples and oranges, (or in this case, grapes and honey). This fundamental difference in starting ingredients shapes everything from brewing to final taste.

Wine is a blend of:

- Grapes

- Yeast

- Sulfur dioxide (for preservation)

Mead combines:

- Honey

- Water

- Yeast

A dash of creativity with additional flavorings, from spices to fruits

The Sweet Influence on Alcohol Content

The sugar content in the primary ingredient is the maestro conducting the alcohol content. Grapes, sugary in their own right, usually bring wines to an alcohol by volume (ABV) of about 12% to 16%. Honey, the star of mead, is a sugar powerhouse. This allows for meads to reach an ABV of up to 22%. But honey's versatility lets you play it down a notch, crafting meads with a more gentle ABV of around 5%.

Tasting Notes: A Symphony of Flavors

When we delve into the realm of taste, mead and wine each conduct a symphony of flavors that are as distinct as their origins. Both beverages can traverse the spectrum from dry to sweet, but that's the beginning of their sensory journey.

Wine, with its rich heritage, offers a robust, wide palette of flavors. Depending on the grape variety and winemaking techniques, you might experience a bold, tannic red that envelops your palate, or a crisp, refreshing white that dances with citrus notes. The aging process, particularly in oak barrels, can introduce nuances of vanilla, spice, or smoky undertones, adding layers of complexity. Each wine reflects its terroir—its unique geographic characteristics of origin, including soil, climate, and topography, lending an almost endless variety to its profiles.

On the other hand, mead sings a different tune. The type of honey used plays a starring role in its flavor composition; honey, by its nature, is a tapestry of the flowers and environments from which it's harvested. This results in meads varying wildly—from sweet and floral, echoing the gentle whisper of clover or lavender, to rich and bold, reminiscent of darker, woodsy tones. When additional ingredients like fruits, spices, or even hops are introduced, the flavor profiles of mead expand even further. A raspberry melomel, for example, can burst with fruity vibrancy, while a spiced metheglin might offer warming notes of cinnamon or nutmeg.

Uncorked Longevity: Mead's Advantage Over Wine

Here's something fascinating about mead: sweeter varieties have a knack for staying fresh longer than wine once they've been opened. It's all thanks to the higher sugar content from the honey, which acts like a natural preservative. Properly stored, an opened bottle of sweet mead can remain enjoyable for an impressively long time, sometimes even years.

Wine, on the other hand, tells a different story. Once opened, most wines prefer a brief spotlight, staying fresh for just a few days before they begin to lose their charm—although, there are exceptions like port or sherry, which have a longer period of grace.

Fermentation and Aging: A Study in Contrast

Both mead and wine undergo fermentation, where sugars are turned into alcohol by yeast. But here's where their paths diverge. Mead often races through fermentation, ready to be enjoyed in weeks or months, yet aging can add layers of complexity to its profile.

The journey of wine is more varied. Depending on its type, it can take a longer path through fermentation and aging. Think of red wines aging gracefully in oak barrels, developing depth and flavors that only time can bestow. The aging potential of wines is a kaleidoscope of possibilities, with some reaching their zenith after many years.

A Journey Through History

Exploring the rich histories of wine and mead unveils a fascinating journey through time and culture. Both of these cherished beverages share roots in ancient times, believed to have sprung from the natural fermentation of wild yeasts. Yet, their stories naturally diverge, painting unique pictures across the global canvas.

Let's first wander through the annals of mead's history. It's a narrative that spans the globe, with mead finding a place in every corner of the world for thousands of years. This widespread presence owes much to the humble bee. Bees, found almost everywhere, have made honey a universally accessible ingredient—unlike the more geographically bound grape. Thus, mead is a global companion to humanity, transcending borders and climates.

However, wine flourished in the Mediterranean, including parts of Africa, Europe, and the Near East. This was mainly due to grape cultivation being limited by specific climatic needs. It wasn't until the wheels of world trade began to turn that wine spread its vines to other parts of the world, gaining recognition and popularity far from its Mediterranean cradle.

The narrative takes an intriguing turn when we look at the shifts in popularity. Despite its widespread availability, mead saw a decline with the advent of distilled spirits in the 1700s and 1800s. The challenge of harvesting honey, as opposed to the relatively simpler processes of making wine or beer, contributed to mead's gradual eclipse. However, the recent resurgence in craft brewing has rekindled interest in this ancient drink. The relaxation of home brewing laws sparked a new curiosity among brewing enthusiasts, leading to a renaissance of mead-making, particularly in the United States and Europe.

In the next part of our book, we'll peer into a detailed chronicle of mead's history, travelling through the centuries to uncover how this ancient beverage has traversed various cultures and periods, maintaining its allure and significance across the globe.

The Legalities of Homebrewing Mead: What You Need to Know

If you're considering stepping into homebrewing, especially in mead, you might be curious about the legal side. Well, you need to be curious!

Homebrewing is an engaging hobby for creating beer, mead, or cider. Imagine the joy of crafting a drink that's uniquely yours, tailored to your tastebuds! It's not just about the final product; it's the journey of mixing, fermenting, and tasting your creation. Whether it's about saving money, experimenting with flavors, or making a healthier drink—homebrewing offers a range of possibilities. And today, it's easier than ever to start, with all the necessary kits and equipment just a click away online.

Now, let's talk legality. It's essential to know the rules of your playground. Each country has its set of regulations regarding homebrewing. These rules ensure safety and compliance with alcohol production standards. In some places, like the United States, brewing mead at home is generally okay, but if you're thinking of distilling spirits—that's a whole different ball game, often requiring specific permits.

Embarking on Homebrewed Mead Adventures in the U.S.

If you are in the United States and dreaming of brewing your mead—Well, you're in luck! Brewing mead at home here is not just a fascinating hobby, but also perfectly legal. Unlike the intricate laws surrounding distilled spirits, mead, (which is non-distilled), enjoys a more relaxed legal status.

Peering into the world of homebrewing is exciting. In many places across the U.S., you'll find a thriving community of homebrewers and a wealth of resources. Imagine walking into a local shop and picking up a kit with everything you need to make mead or beer. It's that easy and accessible!

However, as with any good thing, there are boundaries. If you're 21 or over, you're legally allowed to brew mead at home. But keep in mind there's a cap on how much you can produce—up to 200 gallons per year if two or more adults live in your household. It's important to note that selling your homebrew is off-limits. But what about the silver lining? You can share your homemade mead with friends, or gift it. There's something special about sharing a bottle of your mead—it's a personal touch that just can't be bought.

While federal law is generally supportive of homebrewing, local laws can vary. For instance, certain areas in Alaska have their own rules. It's always a good idea to examine your local regulations to avoid any surprises.

Nationally, homebrewers are under the umbrella of USC Title 26, Subtitle E Ch51, primarily due to alcohol production taxes. So, while brewing for personal enjoyment is excellent, selling your mead could complicate things with tax laws. And a word to the wise: distilling alcohol at home without the necessary paperwork is *a serious no-no*.

Navigating the Legalities of Homebrewing Mead in Europe

Setting out on the mead-making journey in Europe? It's a fascinating endeavor, but navigating various legal landscapes across different countries is essential. Let's explore what this means for aspiring mead-makers across the continent.

Each European country has its own set of rules, creating a diverse set of regulations.

In the U.K., brewing mead at home has the green light. But, if you're considering selling your creation, there are rather complex regulations: you'll need to gear up with the correct licenses—a personal license and a premises license for selling to the public and registration with the Alcohol Wholesalers Registration Scheme (AWRS) for business sales. In addition, you'll need insurance, a registration with the Food Standards Agency, and a license for producing wine. Setting up a private limited company is also a good idea. And remember, tax obligations are part of the package when selling.

In the Czech Republic, homebrewers can create up to 2,000 liters of beer or wine per household each year. But remember, the Customs Office likes to be kept in the loop about your beer-brewing.

Denmark is pretty relaxed about brewing beer for personal use—no caps on the amount. However, distilling your spirits isn't on the table for individuals.

Head over to Finland, and you'll find that brewing mead for personal enjoyment is fine, but leave the spirit distillation to the commercial pros.

Sweden and Spain are on the same page, welcoming homebrewed mead, as long as it's not for sale.

A Peek Beyond Europe

In Australia, brewing mead and wine for yourself is great. Still, distilling spirits calls for a special license and approval from the Australian Taxation Office, especially if you own a large still. Canada offers a bit more regional variation, with provincial laws governing liquor and federal rules handling the taxation and importation aspects.

Transitioning to Commercial Mead Production

If you've been nurturing a dream to take your homebrewed mead to the wider world, here's what you need to know, to *realize* that dream.

First, to brew mead commercially, getting all your legal ducks in a row is crucial. This means gathering the necessary documents and licenses that sanction your operation. And don't forget taxes. Staying on top of your fiscal responsibilities ensures you can operate with peace of mind and legitimacy.

Securing a brewing license is a big step. It's a process that demands time, effort, and a lot of patience. But if your ambition is outgrowing the 200-gallon limit of homebrewing, this is your path forward. Just be prepared for a journey that's as challenging as it is rewarding.

If you're dreaming of turning your mead-making hobby into a thriving business, remember: selling your homebrew without a license is off-limits. And it's not just selling; bartering your homebrew also falls into a legally gray area. Homebrewing laws are pretty straightforward—it's for personal enjoyment, *only*. So, while it might be tempting to trade your mead with fellow enthusiasts, the safest bet is to keep it for personal use, and sharing with friends.

Understanding the Brewing Timeline for Mead

When setting off on your mead-making adventure, you might initially wonder, "How long does it take to brew a batch of mead?" It's a great question, and the answer isn't as straightforward as you think.

Firstly, let's debunk a common myth: not all meads need a full year to mature. Yes, some traditional styles require this long aging process to reach their peak, but numerous recipes promise delightful results in far less time. The key factors influencing your mead's brewing time include its Alcohol By Volume (ABV) and the health of your yeast during fermentation.

ABV is more than just a number in the world of mead—it's a guide to how long your mead should rest before it's at its best. Lower ABV meads generally mature faster. This is because the strong alcoholic flavor common in many brews needs time to mellow out, becoming more palatable. The higher the alcohol content, the longer this process takes. However, lower alcohol levels mean a shorter waiting period.

But here's some encouraging news for those who prefer a stronger mead: expediting the maturation process of high-ABV brews is possible. Achieving this quicker turn-around hinges on effective yeast fermentation and a well-balanced brew. There are techniques to master this balance, which we'll cover in the following chapters. If you're aiming for a potent mead, don't be disheartened; with the right approach, you could savor it in just six months!

CHAPTER 1
Different Styles of Mead

In this chapter, we introduce you to the diverse types of mead, focusing on its three main classifications. These categories are selected for their widespread recognition and importance, paving the way for a good understanding of the vast spectrum of our special beverage, mead. We'll look at the distinctive features of each type, covering everything from the selection of ingredients, to the nuances of aging and sweetness. This section aims to enlighten beginner readers with a comprehensive introduction, while providing enriching details for the more experienced mead aficionado.

Classification of Mead based on Ingredients

Traditional Mead

Glass of Mead/Credit: Brent Hofacker (www.Shutters tock.com)

Traditional mead, often also called "show mead", stands out in the mead-making world for its striking simplicity. This type of mead is crafted from just three ingredients: honey, water, and yeast. The beauty of traditional mead lies in this simplicity, shining the spotlight directly on the honey used. It's like a blank page, where the quality of the honey is not just an ingredient, but the story's central character.

The sweetness level of traditional mead is a critical defining feature, ranging from dry to sweet. This isn't just about flavor; the residual sugars left after fermentation influence the body, or viscosity, of the mead. Sweeter meads typically have a richer, more luxurious mouthfeel, showcasing how fermentation can transform simple ingredients into a complex, delightful beverage.

A lively debate exists within the mead-making community regarding the purity of traditional mead. While some mead-makers hold a purist view, advocating for a composition of only honey, yeast, and water, others refer to historical practices to support additional flavorings, albeit in small amounts.

Brewing traditional mead is often seen as the pinnacle of a mead-maker's craft. The absence of additional flavors to mask imperfections means that every aspect of the mead—the fermentation process, the yeast's health, and the honey's quality—must be flawless. Making a great traditional mead is a testament to the mead-maker's skill, turning each batch into a reflection of their expertise—and of course the unique qualities of the honey used.

Acerglyn

Maple syrup bottles/Credit: Cindy Creighton (www.Sh utterstock.com)

Acerglyn, a delightful twist in the mead-making journey, blends traditional honey sweetness with the rich, earthy essence of maple syrup.

In the world of mead-making, acerglyn brings a unique twist that sometimes inspires passionate discussion among enthusiasts. The crux of this discussion? Its use of maple syrup. For traditionalists in the mead community, the essence of mead is deeply infused with honey. These purists often argue that true mead should be predominantly honey-based. Acerglyn, with its substantial use of maple syrup, often finds itself at the centre of this debate, as some question whether it fully qualifies as a mead!

The creation of acerglyn starts with a crucial decision—the choice of maple syrup. Here's where your role as mead-maker becomes exciting. Do you envision a subtle maple undertone? Or a bold, maple-forward character? The ratio of syrup to honey is your trump card, allowing you to craft an acerglyn that resonates with your palate´s preferences.

Fermentation is the stage where patience becomes your best ally. It's not just about letting the yeast do its work; it's about allowing the unique combination of honey and maple syrup to mature and develop. The conditions you maintain during fermentation—the temperature, yeast strain, and time—are critical in shaping the acerglyn's flavor.

There's a seasonal aspect to acerglyn that adds to its charm. Its warm, comforting flavours are often associated with autumn and winter, making it a perfect companion for cooler days. The traditional maple syrup harvesting time in these seasons adds a touch of nostalgia and authenticity to each batch of acerglyn brew.

Melomel

Melomel is a vibrant player in the team of mead, where honey's natural sweetness meets diverse fruit flavours. In making melomel, the choice of fruit plays a central role; from the tartness of berries to the mellow sweetness of stone fruits, each type of fruit brings its unique character to the mead.

The beauty of melomel lies in its balance. It's a delicate alignment of the distinct flavors of fruit and the natural characteristics of honey. This balance doesn't demand that fruit or honey dominate; it calls for a blend where both can shine through. The skill in crafting melomel comes from adjusting this balance to suit your palate—a process that is rewarding and challenging.

Melomel's versatility is found in its various subcategories, often named following historic English conventions. For example, pyment, a specific type of melomel, blends grapes with honey—merging the art of winemaking with mead-making! More on this soon! This diversity makes melomel one of the most popular styles among mead-makers, offering pretty much endless opportunities for interesting flavor combinations.

Searching the history of melomel, we find its roots growing alongside the ancient practice of mead-making. Historical records, like those of Pliny the Elder in his *Naturalis Historia,* suggest that the preservation of fruits in mead was common, utilizing the alcohol content of mead to extend the fruits' longevity. This practice hints at the ancient origins of melomel—possibly as old as mead itself.

Next, we will travel into the world of melomel, focusing on some of the most common and renowned types.

Cyser (Apple Mead)

Apple mead/Credit: nobito (www.Shutterstock.com)

Cyser, a charming variety within the melomel family, is a delightful fusion of honey and apples, creating a mead rich in tradition and flavor. Unlike traditional apple cider that relies on sugar, cyser uses honey, bringing a special floral nuance that marries beautifully with the natural taste of apples. Cyser demands careful blending in which the sweetness of honey and the character of apples harmonize perfectly.

The art of crafting cyser lies in balancing the natural acids and tannins from the apples with the sweetness of the honey. This balance determines whether the cyser will lean towards a dry or sweet profile. The choice of apples plays a significant role here. While most commercial apples are bred for eating, traditional cider apples, (often harder to source), can elevate a cyser to new heights of complexity and flavor. Sometimes, homebrewers may add acid and tannin to adjust the taste, but the most authentic and intricate flavors typically come from the natural properties of the chosen apples.

We find the history of cyser steeped in centuries of tradition. Mentioned in ancient texts and folklore, cyser was particularly revered during the Middle Ages, especially in regions where apples were abundant. Often associated with autumnal festivities, it was a celebratory drink that marked the harvest season. Today, it continues to be a prized beverage among craft brewing enthusiasts, cherished for its unique taste and historical significance.

The brewing process of cyser begins with carefully selecting high-quality honey and apple juice. These ingredients are combined and heated to pasteurize the mixture. Once cooled, it's transferred to a fermentation vessel, and yeast is added. The fermentation, taking several weeks, is where the magic happens, as honey and apple flavors meld and mature. After fermentation, the cyser is bottled and aged, developing its full flavor profile.

The taste of cyser is a delightful mix of honey's sweetness with the tartness of apples. Depending on the honey-to-apple ratio and the type of yeast used, the final product can range from dry to sweet—each batch reflecting the brewer's skill in that regard.

Pyment (Grape Mead)

As mentioned earlier, pyment, another captivating variant within the melomel family, merges the grapes' essence with the rich mead-making tradition beautifully. This grape mead is most commonly crafted by fermenting honey with grape juice, though some mead-makers take a creative approach by sweetening homemade grape wine with honey, or blending grape wine with mead post-fermentation.

The artistry in creating a memorable pyment lies in achieving a harmonious blend, where the sweetness, acidity, tannin, and alcohol complement both the honey and grape aspects. Here, the grape selection is crucial. While wine grapes are often preferred for their rich flavors, a wide array of wine *and* table grapes can be used, each bringing its unique character to the mead.

What makes pyment particularly fascinating is the influence of the grape's yearly variations due to changing growing conditions. These fluctuations in rainfall and temperature mean that the expected aromas and flavors differ annually, even within the same grape variety.

Tracing back to the origins of pyment, we find ourselves lost in the annals of history. It's a drink that likely predates recorded history, with ancient Egyptian, Greek, and Roman cultures referencing wines made or sweetened with honey. The word "pyment" hails from the Middle Ages, with historical beverages like the Roman "mulsum" and Greek "melitites" resembling what we know today as pyment.

Berry Mead

Berry mead, another delightful member of the melomel family, offers a tapestry of flavors, uniting those little colorful fruits—berries—with the timeless art of mead-making. This type of mead celebrates berries in their varied forms—from the familiar raspberries and strawberries to the less common lingonberries and elderberries.

Berry mead creation is a careful blend of art and science. It involves the fermentation of honey with a selection of berries, be it a single type or a mix. The charm of berry mead lies in how the distinct flavors of the berries meld with honey. The flavor profile can vary quite significantly depending on the berries chosen, with the spectrum ranging from subtle and fruity to richly intense.

Aroma and of course flavor are central to the character of berry mead. The aroma can vary from a soft, honeyed fragrance to a bold, fruit-forward bouquet, depending on the honey and berries used. The fruit character should be authentic and well-integrated, complementing (rather than dominating) the mead. The honey, too, should be noticeable, its sweetness harmonizing with the fruit's natural acidity and tannin levels.

When it comes to appearance, berry mead presents a stunning array of colors, influenced by choice of fruits and honey. The colors are often lighter than the fruit, adding a delicate hue to the mead. The mouthfeel usually resembles wine, with natural acidity and tannin from the fruit contributing to the body and balance.

In crafting berry mead, the goal is to achieve a delightful harmony between the fruit and honey. Whether the mead is dry or sweet, the balance of tannin and sweetness is key, especially with darker berries like blackberries that can impart a wine-like tannin presence. We will briefly cover the two most famous types of berry mead, the bilbemel, black mead, and the morat, each offering its distinct and captivating profile.

Bilbemel

Bilbemel mead offers a unique berry-infused experience, traditionally crafted with bilberries but equally enchanting when made with blueberries.

In crafting bilbemel, choosing between bilberries and blueberries isn't just a matter of availability and taste preference. Bilberries bring a robust, tart flavor, lending the mead a deep, earthy character. Blueberries, in contrast, offer a milder, sweeter berry note. This choice influences the taste and color of the bilbemel, ranging from a deeper, almost mystical hue with bilberries to a brighter, more radiant color with blueberries.

The art of making bilbemel is balancing honey's sweetness with the berries' natural tartness. Whether you choose the intense flavor of bilberries or the gentler taste of blueberries, the key is to find harmony between these elements.

Black mead (blackcurrant mead)

Blackcurrant mead, affectionately known as black mead, is a rich and flavorful addition to the mead family, celebrated for its deep, vibrant flavors. This mead variety, infused with the essence of blackcurrants, is not just a drink but a sensory experience, combining the tartness of these berries with the sweetness of honey in a beautifully balanced way.

The heart of blackcurrant mead is its bold use of those little dark fruits—blackcurrants. These berries, known for their intense flavor and tartness, impart a depth to the mead that's both intriguing and satisfying. The striking, deep purple color adds to the allure, making it as visually appealing as it is delicious.

The flavor profile of blackcurrant mead is particularly noteworthy. The natural tartness of the berries is beautifully offset by the honey's sweetness, resulting in a complex and layered mead. This makes it an excellent pairing for various foods, especially those that benefit from fruity acidity, such as rich meats and cheeses. For culinary enthusiasts and mead lovers, blackcurrant mead opens up a world of pairing possibilities.

Morat (mulberry mead)

Morat, a special kind of mead, captures the essence of summer in its rich, flavorful composition, made distinctively with fresh mulberries. Typically crafted during the summer (or late summer), this is the season when mulberries are at their most flavorful—bursting with juicy goodness. The appeal of morat lies in its delightful color and taste, a delicious blend of sweet berry notes intertwined with a honey wine flavor, evoking images of sunny summer days.

What sets morat apart is not just its lovely flavor but also its readiness for enjoyment. Morat reaches its prime relatively quickly, unlike other meads that might require a longer maturation period. This makes it an excellent choice for those eager to enjoy their craft without the prolonged waiting period often associated with mead-making. It's particularly appealing for those new to mead-making, or anyone who loves the idea of a mead that doesn't ask for months of patience. Morat, with its easygoing nature and delightful taste, is like a celebration of summer—ready to be enjoyed in its youthful vibrancy.

Metheglin (spiced mead)

Glass of Metheglin/Credit: Aleksandra Berzhets (ww w.Shutterstock.com)

Metheglin, a distinctive form of mead, is characterized by its delightful blend of spices with honey.

Tracing the name "metheglin" to its roots, we find ourselves touching upon the Welsh language. The word originates from "meddyglyn", combining "meddyg"—meaning "doctor" or "healer" (a term derived from the Latin "medicus")—and "lyn", translating to "liquor". Thus, metheglin historically alludes to "healing liquor", a nod to its past as a medicinal remedy. Interestingly, the Welsh word "meddyg" also forms the foundation for the English word "mead", linking this spiced mead to its linguistic heritage.

Historically, metheglin's significance extends beyond that of a mere beverage. It was integral to folk medicine in Wales, traditionally used to treat colds and other ailments. Its reputation as a healing drink even reached the Romans, with Julius Caesar and his legions recorded to have enjoyed metheglin during their campaigns in the United Kingdom.

The variety and complexity of metheglin recipes flourished in the Middle Ages. However, due to the high cost of spices at the time, metheglin was a luxury typically reserved for the upper echelons of society—the royals and the elite.

The essence of metheglin lies in how the spices are integrated into the mead's core elements—honey, acid, tannin, and alcohol. The choice of spices in metheglin will significantly influence its character, leading to a varied range of unique final products. Crafting metheglin involves striking a balance, where the spices are distinct—yet blend seamlessly with the mead's base flavors.

In terms of aroma, metheglin can range from subtly fragrant to richly aromatic, depending on the mead's sweetness and the spices used. The character of the spices should be clean and natural, with some, (like ginger or cinnamon), offering more robust aromas than others, (like chamomile or lavender). This diverse aromatic profile should complement, not overpower, the natural fragrance of the honey. If a specific variety of honey is used, its unique aromatic qualities should also shine through, adding depth to the mead's aroma.

Visually, metheglin typically retains the standard appearance of mead, with the color largely unaffected by the spices. The flavor profile of metheglin is where the magic of spices comes into full play. In blends of multiple spices, achieving a harmonious balance is crucial. The mouthfeel in metheglin is generally akin to traditional mead, but including certain spices can introduce elements like a hint of astringency, or a warming sensation. These additional sensory features should be well-balanced and enhance the overall drinking experience. It's important to note that mead is categorized differently when spices and other ingredients like fruits or cider are used.

Rhodomel (rose mead)

Rhodomel and rose hips/Credit: Flegere (www.Shutte rstock.com)

Rhodomel, a distinctive variant within the mead family, is celebrated for its graceful infusion of rose elements with honey.

The heart of rhodomel is its harmony of the essence of roses with honey, which can be achieved using various parts of the rose, such as petals and hips, each lending a unique facet of the rose's character to the mead. The challenge in crafting rhodomel is to achieve a balance where the rose's presence is *perceptible* yet subtle, *enhancing* rather than overshadowing the honey's natural flavors.

In rhodomel, the aroma is typically the gentle interplay of rose and honey, creating an inviting and complex scent. The flavor should follow suit, presenting a delicate blend where the rose's floral notes complement the mead's sweetness and body. This integration should feel seamless, offering a smooth and pleasant taste without the rose character becoming too dominant, or too much like perfume.

Capsicumel

Capsicumel, a distinctive variant in the diverse world of mead, is renowned for its bold incorporation of... wait for it... chili peppers! This style differentiates itself from the traditional metheglin, which typically uses sweet spices and herbs. Capsicumel's hallmark is the integration of peppery heat, adding a fiery dimension to the mead.

The core aim of capsicumel is to capture the intriguing interplay between the warmth of the alcohol and the heat from the chilies. This sensation of heat is, of course, subjective and varies greatly depending on individual preferences. Some might delight in a pronounced burning sensation, while others favor a gentler warmth. The art of crafting capsicumel lies in finding that sweet spot where the heat complements, rather than overwhelms, the mead's natural flavors.

The flavor profile of capsicumel can range from intensely spicy to subtly nuanced. Interestingly, meads with a sweeter profile often pair well with spicy elements, as the chili's heat can reduce the perception of sweetness. This interaction allows other flavors within the mead to become more pronounced. In dryer meads, where less sugar is present due to fermentation, adding chili heat can enhance the honey's natural flavors, adding a layer of complexity to the drink.

Bochet

Bochet, a mead style with medieval French origins, has recently found a renewed following in the homebrewing community. This distinctive mead, known for its caramelized honey base, offers a unique taste experience that stands out in the world of mead-making. The modern interpretation of bochet is a delightful flavor exploration, with caramelization at its heart, and the option to infuse spices.

The resurgence of interest in bochet can be traced back to the rediscovery of an ancient recipe from 14th-century Paris. The publication of *The Good Wife's Guide: a Medieval Household Book* by Cornell University Press in 2009, which included a translation of *Le Ménagier de Paris*, unveiled this historical mead recipe to a broader audience. The term "bochet", derived from the original French recipe, has become synonymous with this style of mead made from caramelized honey.

Creating bochet is an art that revolves around the caramelization of honey. This crucial step involves carefully boiling the honey until it reaches the desired level of caramelization, significantly shaping the mead's flavor and aroma. This process intensifies the honey's sweetness and gives the mead a rich, complex character. Additionally, caramelizing the honey reduces its water content, potentially leading to a higher alcohol content and a more syrupy texture in the finished mead.

Depending on the level of caramelization, the color of bochets may vary from a deep, almost black, to a rich amber. Aging bochet on oak or other types of wood can add further depth to its flavor profile. Many brewers also experiment with adding spices or fruits to enhance their character, enriching the mead's complexity. Bochets are celebrated for their deep, multifaceted flavors and aromas, often enjoyed as a luxurious treat or a soothing digestif.

Bochetomel

Bochetomel is a masterful blend. This mead style artfully merges the rich, caramelized depth of a bochet with the vibrant fruitiness of a melomel. Its fusion highlights the complex, toasty notes from caramelized honey, skilfully paired with various fruits' refreshing, tangy flavors.

In crafting bochetomel, caramelizing honey, (a technique central to bochet), creates a warm, rich base. This is then beautifully complemented by adding fruits, (a key element in melomel), introducing layers of sweetness and acidity. The result is a mead that speaks volumes in terms of flavor; it's a delicious blend of deep, almost smokey tones from the caramelized honey with the lively, fresh notes of the fruits.

Braggot

Braggot/Credit:trevmo64(www.istockphoto.com)

Braggot, a distinctive and harmonious mix of mead and beer, uniquely combines the rich traditions of both brewing worlds. Typically involving grains, (often malted barley), this hybrid beverage also embraces *other* grains, each contributing to its diverse flavor profile. Crucial to its creation is the malting and mashing process, reminiscent of beer-making, where complex starches in grains are converted into sugars for yeast to ferment.

A notable aspect of Braggot is the optional use of hops, setting it apart from modern beers. This flexibility allows the grain and honey aspects to range from subtle nuances to prominent flavors, without the hops overpowering the balance. Often found in a balanced mix, braggot can vary in its honey-to-malt ratio, offering a range of flavors and textures. Braggot's identity is as intriguing as its taste. Some enthusiasts argue it's essentially a *beer* enriched with the sweetness of mead; while others see it as a grain-infused *mead*.

While modern beers are known for their hops, in braggot, hops take a back seat, making their presence optional and never overwhelming. The true essence of a braggot lies in its balance, showcasing an equal harmony of honey's sweetness and the richness of grains.

The aroma of a braggot varies, influenced by factors like sweetness, strength, and the choice of beer base. It should offer a nuanced blend of honey and beer or malt, with the potential for a noticeable varietal character if a specific honey type is used. Similarly, the beer style or malt type can also imprint its character on the aroma. Hop aroma, if present, should integrate seamlessly with the other elements. Braggot's appearance departs from typical mead, leaning more towards beer, with clarity ranging from good to brilliant.

The flavor in braggot is in between beer and mead, with the intensity shaped by sweetness, strength, and the base beer style. The malt character ranges from simple base malts to complex caramel, toast, or roast flavors. If hops are used, they should complement the overall flavor profile, adding another layer to the braggot's complexity.

Regarding mouthfeel, braggot diverges from standard mead, offering a smooth texture without astringency. The body can vary, influenced by the sweetness and strength of the brew. Carbonation levels in braggot can also vary, adding to its diverse profile.

When crafting a braggot, choosing the honey and malt is critical, each contributing to the final flavor and aroma. The aim is to create a beverage where the qualities of honey and beer are perceptible, yet harmonious, neither overshadowing the other. Sometimes also known as "bracket", braggot's fermentable sugars are a balance of malt or malt extract and honey, leaving room for creative interpretation by the brewer.

Coffee Mead

Coffee mead is a fascinating blend where the rich world of coffee meets the age-old tradition of honey wine. This innovative concoction infuses the essence of coffee beans into honey wine, offering a harmonious interplay of such different flavors. The coffee beans complement the yeast's natural flavor profile in the mead, and elevate the drink to a new level of deliciousness!

There are various methods for infusing the coffee into mead. Still, the most effective and favored ones involve using cold brew, or adding whole coffee beans directly into the fermenter. These techniques are renowned for extracting a purer, more pronounced coffee flavor, enriching the mead's taste.

One of the most intriguing aspects of coffee mead is its incredible versatility. It can range from sweet to dry, light to dark, and can be crafted solely with honey—or incorporate other ingredients. This flexibility allows for various flavor profiles, catering to diverse tastes and preferences. Coffee mead can also vary in its presentation—it can be carbonated or still, and its alcohol content can range from giving a gentle "buzz" to a more robust, intense effect.

Not only is coffee mead adaptable in taste and composition, but it's also versatile in how it's served. It can be a refreshing, ice-cold beverage perfect for summer days; or a warm, comforting drink for chilly winter nights. Beyond its delightful taste, coffee mead marries the energizing qualities of coffee with the antioxidant-rich properties of honey, making it an enjoyable—and beneficial—beverage.

Hippocras

In mead-making circles, hippocras is recognized as a spiced or herbed pyment. However, it's important to note that, historically, *true* Hippocras is not a mead, but a type of mulled wine sweetened with honey. This beverage has its roots in ancient medicinal practices, where a variety of herbs were steeped in sweetened wine for a day, and then filtered through a conical cloth filter, known as a *manicum hippocraticum* or Hippocratic sleeve (originally devised by Hippocrates, the famed 5th century B.C., Greek physician as a filtering method for purifying water).

Unlike a metheglin, where herbs and spices are integrated into the must and impart their flavors during brewing, most modern mead hippocras recipes involve steeping herbs and spices in a pyment for 24 hours. After this infusion period, the mixture is then filtered and bottled. This method results in a drink that captures the essence of the spices and herbs *without* having them exist in the final product.

Oxymel mead

Oxymel mead, an elixir steeped in history, artfully combines the sweet richness of honey with the sharpness of vinegar. This ancient concoction, dating back to the era of Hippocrates, was initially crafted for its health-promoting properties. Derived from the Greek "oxy," meaning acid, and "mel," meaning honey, oxymel mead encapsulates its core ingredients in its name.

This traditional blend primarily features honey, vinegar, and water, each contributing to its distinctively balanced sweet and sour profile. Historically valued for its medicinal benefits, today's oxymel mead has evolved into a beverage cherished for its unique taste and versatility.

The flavor of oxymel mead will intrigue you with its complexity, where the sweetness of honey melds seamlessly with the tartness of vinegar, creating a refreshingly layered drink. Its ability to cleanse the palate makes it an excellent choice as an aperitif, or as a complement to a meal.

Modern interpretations of oxymel mead often add an array of herbs and spices, enriching its flavor and healthful properties. Its adaptability shines, whether sipped on its own, mixed into creative cocktails, or used in culinary ways, like in dressings and marinades.

Mulled mead

Mulled mead, with its origins in the 2nd century A.D., is a testament to the adaptability and ingenuity of ancient cultures. The Romans, renowned for their love of wine and known to face the harsh winters of central and northern Europe, pioneered mulling—warming up alcoholic beverages and infusing them with spices. This innovation was born out of necessity, a way to cope with the stark contrast between the mild Italian climate and the bitter cold of regions, like central Germany and Scotland's borders.

In their quest for warmth and comfort, the Romans began heating the local wines they produced and enriching them with various herbs and spices. These additions were not just for flavor; they were chosen for their medicinal properties and the extra warmth they brought, a crucial feature in the frigid European winters. This practice quickly resonated with the local populations under Roman rule, who embraced and expanded upon it. Soon, mulling was applied to various beverages, including beers and ciders in Roman Gaul (modern-day France) and England, and eventually, mead in the northern territories.

This adaptation led to mulled mead, a warm, spiced version of the traditional honey wine. Despite the decline in general mead consumption over the centuries, mulled mead has endured in certain cultures, particularly as a part of traditional celebrations. It remains a beloved part of Yule and Christmas festivities in Sweden and Norway, and a staple in Finland's May Day celebrations.

Mulled mead, a delightful winter beverage, embodies the tradition of mead-making with a twist of aromatic mulling spices, offering comfort during the colder months. This beverage is *more* than just warmed honey wine; it blends cinnamon, cloves, allspice, and nutmeg, each lending warmth and distinct flavors. Adding citrus elements, like orange peel, further enriches its taste.

The essence of mulled mead is found in the gentle process of heating it and steeping it with these spices. This method allows the spices to release their essential oils, marrying beautifully with the sweetness of the honey. The result is a soothing, richly spiced drink, perfect for sipping on a chilly evening or as a festive treat during holiday gatherings.

As the mead warms and the spices infuse, the aroma released is as inviting as the drink, creating an ambiance of warmth and comfort. Simple to prepare yet deeply rewarding, mulled mead can be tailored to individual tastes, allowing for customization in the choice and balance of spices.

One finds on this journey of mead discovery that ageing plays a pivotal role in character development. Let me present a classification of meads—distinguished by how time crafts their essence.

Short Mead Versus Long Mead

Short mead is the sprinter in the race against time. It's crafted for those keen to dive into mead's delights—without the long wait! You could be sipping on short mead in just a few weeks to a few months after fermentation, enjoying a light, refreshing drink, perfect for a laidback gathering. It's the kind of mead that doesn't require loads of your patience, offering a vibrant taste of honey, just right for an impromptu toast or a sunny afternoon.

On the other end of the spectrum you'll find long mead, the marathon runner, ageing gracefully—sometimes over many years. It's the choice for the sentimental, those who see time as an ally in crafting a drink with character. Long mead develops complexity and richness worth the wait, as months flow into years. With each passing year, it becomes smoother, bolder, and more dignified, turning into a beverage that's saved for life's milestones, to be savored slowly, sip by sip.

Classification of Mead Styles by Alcohol Content

One discovers that the alcoholic strength of this venerable drink varies as much as its flavors. This variance stems from the amount of honey and other sweet additions whisked into the mix before the magic of fermentation takes its course.

You have the delightfully light hydromel at one end, tipping the scales with an alcohol content gently nestled between 3.5% and 7.5%. It's the go-to for a breezy afternoon—a mead that whispers rather than shouts, carrying a subtle hint of honey—just right for toasting life's small victories.

Then there's the standard mead, striding confidently in the middle with an alcohol presence of 7.5% to 14%. It's the mead that strikes a perfect chord between robust and refined, presenting a fuller taste of honey that's rich, but not overbearing—the kind of drink that's as at home at a dinner party as it is at a casual get-together.

The bold and indulgent sack mead sits at the other end of the spectrum. With an alcohol content that ranges from a hearty 14% to an impressive 18%, it's mead with gravitas. This is the one you reserve for those special moments, offering a depth of flavor and a warming sensation that only comes from a mead that's not afraid to flex its strength.

The essence of alcohol in mead comes down to ethanol—that colorless, potent spirit that carries the aromas and flavors of the mead. It can introduce a touch of bitterness, sometimes perceived as the counterpoint of sweetness, and it's known for that comforting warmth it brings to the throat.

Crafting the perfect mead involves balance and precision. The mead maker's skill lies in adjusting the honey, considering each addition to create the desired strength. A vigorous fermentation, coaxed by the selection of yeast and the watchful eye of the brewer, can yield a higher alcohol content. In contrast, a gentle fermentation might keep things lighter.

Classification Based on Sweetness

In mead-making, the sweetness level is a crucial aspect that defines each mead's unique character. This sweetness, primarily derived from honey, might also come from fruits or other additions in the aforementioned various styles of mead. It's important to remember that sweetness in mead isn't just about taste; it subtly influences both aroma and body, adding to the mead's overall sensory appeal.

Let's explore the different categories of mead based on sweetness:

Dry mead: This kind of mead is distinguished by its low level of sweetness. It's not necessarily completely *devoid* of sweetness, but it is subtle enough to be noticeable. Dry mead is often misunderstood as being completely lacking in flavor. However, it can still possess a certain fruitiness and body, contributing to a nuanced drinking experience.

Semi-sweet (or off-dry) mead: Sitting comfortably in the middle of the sweetness spectrum, semi-sweet meads offer a sweetness that doesn't dominate the flavor profile. This category is about balance, where the sweetness complements the mead's other flavors without overwhelming them.

Sweet mead: Sweetness takes center stage in sweet meads but is carefully managed to avoid being overly cloying. These meads offer a richness and depth of flavor, where the sweetness enhances rather than masks the mead's character. It's a fine line to tread, ensuring the mead remains sophisticated, and doesn't simply devolve into an overly sugary drink!

The process of achieving these varying sweetness levels can be pretty intricate. Sweetness might result naturally from the fermentation process, or be the outcome of intentionally stopping fermentation early. Another method involves adding honey or sugary adjuncts *after* fermentation. However, caution is necessary here to prevent re-fermentation if the mead hasn't been stabilized.

It's also interesting to note that sweetness in mead doesn't correlate directly with its alcohol content. You could have a mead that's light in alcohol yet lusciously sweet, or strong and robust but leans more towards the dry end of the spectrum.

CHAPTER 2

Honey, Water, Yeast: Building Blocks of Traditional Mead

I n this new chapter, we focus on the fundamental ingredients of traditional or show mead, a style loved for its purity and depth. We'll closely examine the roles of honey, yeast, and water—the trio that forms the heart of this classic mead variant. This chapter aims to equip you with the knowledge to appreciate and craft traditional mead, emphasizing its special character shaped by these simple yet profound ingredients. As discussed previously, the ingredients for other mead varieties and beginner-friendly recipes will be detailed in an upcoming chapter, enriching your mead-making knowledge with a broader array of flavors and techniques.

Before we move forward, it's essential to grasp the concept of "must," especially for those just beginning mead-making. "Must" is the term used to describe the foundational mixture of honey and water that forms the basis of your mead. This mixture becomes the canvas for your creativity—where the magic begins. By introducing yeast to the must through a process known as "pitching," you set the stage for a remarkable transformation. Over the following weeks, the yeast works diligently, converting the must into the enchanting beverage known as *mead*, often referred to as "liquid gold" for its delightful, precious result.

Honey: The Sweet Foundation of Mead

Different honey varieties/Credit: Alessandro Cristiano
(www.Shutterstock.com)

Honey, the backbone of mead-making, is a fascinating blend of glucose (38%), fructose (30%), and smaller amounts of maltose (7%). This composition gives it a sweetness level akin to table sugar. With a water content of about 17% and gluconic acid as its primary acid, honey's pH ranges from 3.4 to 6.1. It's not just sweet; honey also contains proteins, complex carbohydrates, trace elements, and minerals.

Honey's natural antimicrobial properties arise from its sugar concentration and low pH, making it hostile to microorganisms. These properties were recognized by ancient Egyptians, who used it as an antiseptic and antibiotic. However, honey's inability to kill dormant bacterial endospores means it's unsuitable for application with infants.

Crafting Mead: Choosing the Right Honey

When embarking on mead-making, your choice of honey is pivotal. It's not just a sweetener but the heart of your mead's flavor. Navigating the honey options may seem daunting. Each brings flair to your mead, from the delicate floral varieties to the deep, robust ones. The honey's color, aroma, and flavor are vital descriptors—but the mead's final taste might differ.

Interestingly, honey from fruit blossoms doesn't mirror the fruit's flavor. In crafting your mead, understanding these nuances in honey selection is vital. This part of the book aims to guide you through the myriad of honey types, focusing on essential factors to consider, ensuring your mead is as special as its maker (you!).

Raw vs Pasteurized Honey

Choosing between raw and pasteurized honey is an important decision. Raw honey, not exposed to heat during processing, retains all its natural enzymes, vitamins, minerals, and its full array of aromas, flavors, and colors, all essential for crafting high-quality, flavorful homemade meads. On the other hand, pasteurized honey, treated at high temperatures, loses some beneficial compounds—but is easier to dissolve in liquids like water. Using local raw honey is often recommended, as it ensures knowledge of its origin and processing methods, enhancing the authenticity of your mead. However, if local honey isn't accessible, imported raw honey is a good alternative, as long as it hasn't been heat-treated during processing. The quality of the honey is fundamental in determining the quality of the mead; good honey will elevate your mead, while subpar honey will limit its potential.

Exploring the Colors of Honey in Mead-making

In the intricate art of mead-making, it's easy to be captivated by the vibrant colours of honey, but let's take a moment to explore *beyond* the surface. While the color of honey is indeed eye-catching, it doesn't tell the whole story about its quality. Moisture content and flavor, critical factors in a honey's quality, remain hidden beneath its colored veil.

Imagine a spectrum of honey hues, beautifully laid out by the USDA (U.S. Department of Agriculture), ranging from the almost clear "water white" to the rich "dark amber":

- water white
- extra white
- white
- extra light amber
- light amber
- dark amber

These colors are not just for show; they're like windows into the honey's essence. Each shade results from a unique blend of minerals and compounds that mix with the sugar, variations based on the floral sources the bees have foraged from. A little-known fact: the darker the honey, the richer it is in these compounds, often brimming with beneficial properties.

When discussing flavor, the color of honey can be a subtle guide. Lighter honey usually has a softer, more delicate taste—ideal for those who love a gentler touch in their mead. Darker honey, contrastingly, often boasts bold and robust flavors. But nature loves to surprise us! Take basswood honey, light yet bold in flavor, or the darker tulip honey known for its mildness.

The timing of the honey harvest is crucial, influencing the final flavor and aroma profile.

Honey isn't just another product; it's a moment captured, influenced by the variety of plants blooming at the time of collection. As honey matures in the hive, exposed to air, it absorbs wild yeasts and other environmental particles. Additionally, external factors like humidity and temperature play a significant role in shaping its final character.

For the discerning mead-maker, early-season honey is often a treasure. Typically lighter in color, with a milder aroma and fewer wild yeasts—a result of lower yeast concentrations in the air during early summer—it's not just about the lighter shade. Early-season honey is often celebrated for its exceptional floral qualities, making it a fantastic choice for crafting top-tier mead.

Understanding Honey Grades in Mead-Making

Really knowing the quality of your honey is essential in the time-honored craft of mead-making. Did you know that the USDA breaks down honey into four specific grades? These grades are based on several crucial factors: moisture content, flavor, the absence of defects, and overall clarity. For a mead-maker, this grading system is an invaluable tool. It guides you in choosing the perfect honey, ensuring your mead turns out just right.

Let's talk about moisture. For honey to meet the standards of USDA Grades A and B, it can't have more than 18.6% moisture. This is pretty important. Why? Because lower moisture content helps keep those wild yeasts and other natural organisms in check. If these little guys get too active in honey with a higher water content, they can mess with the honey's clarity and flavor, and that's *not* what you want in your mead.

Then there's Grade C honey. This one's a bit of a gamble for mead-makers since it contains up to 20% water. This Grade has a higher chance of fermentation, especially if the moisture content exceeds 19%. Fermentation can change the quality of honey, affecting its taste and texture in ways that might not be *ideal* for your mead. Typically, you'll find Grade C honey in commercial food processing, not in a top-notch mead. So, even if it seems like a great deal, it's better to pass it up for a better-quality option.

Flavor is another big deal. Experts look for anything that *shouldn't* be there when honey's graded. Things like smoke trace from bee collection, signs of fermentation, chemical residues, or other odd flavors, including caramelization. Grade A honey is top-of-the-line stuff—free from all these unwanted elements. Grades B and C are considered "practically free" and "reasonably free" from caramelization but are still completely clear of other off-flavors and defects.

And don't forget about the appearance! No one wants honey with bits of comb or propolis floating around in it. That's why checking for visible defects is a big part of grading. Then there's clarity—a clear sign of quality. Grades A, B, and C are ranked as "clear"; "reasonably clear"; and "fairly clear." This is determined by looking for bubbles, pollen grains, and other particles suspended in the honey.

For the dedicated mead-maker, I advise sticking with Grade A (Fancy) or Grade B (Choice) honey. These are your best bets for crafting a mead that stands out in flavor and clarity.

Exploring Varietal Honeys

Each varietal honey is sourced from specific nectar and offers a unique palette of characteristics. This honey, nurtured by beekeepers who strategically place new combs in hives close to chosen crops, results in "single source" honey. Each variety captures the essence of the blossoms or fruits they come from. Take raspberry and blueberry honey, for instance. They don't just share the fruits' names; they also reflect their distinct scents and tastes. This natural mimicry adds a layer of depth and authenticity to your mead.

Honey and pH

Another critical factor is the pH of honey. It directly influences the fermentation process. Generally, starting with a pH above 3.6 is best, though you can adjust if your honey's pH is lower. The total acid content also dramatically affects the mead's flavor. And let's not forget the minerals and ash content in honey. These contribute essential nutrients and nitrogen needed for fermentation. If your honey is low in these elements, you might need to add extra nutrients to achieve the perfect fermentation balance.

It's honey time! Now let's dive into an exciting exploration of the most popular and recommended varietal honeys for mead-making. We'll uncover what makes each unique and why they're the go-to choices for crafting exceptional mead.

Alfalfa

Alfalfa honey is something unique in the mead-maker's pantry. Harvested from the purple blooms that grace the fields of the Midwest, Canadian Great Plains, and the Western U.S., this honey is a treasure waiting to be transformed into liquid gold.

When you pour alfalfa honey, it's a bit like capturing sunlight in a bottle; its color ranges from the purest white to a gentle extra light amber. This honey glows in the brew as you craft your mead, leaving a pale white to a straw-coloured hue as inviting as a dawn sky.

The scent? Imagine a walk through a field early in the morning, the aroma of beeswax and fresh hay rising with each step. That's the aroma that alfalfa honey exudes—unassuming yet memorable.

And the taste? It's as if that same tranquil walk yielded a flavor—mild, light, and as delicate as a breeze.

This is why it's perfect for traditional meads. Alfalfa honey doesn't overpower the brew; it subtly enhances, making it ideal for those who want their mead to be a nod to tradition rather than a departure. It's important not to mistake its quiet nature for simplicity, though. Alfalfa honey can be mistaken for clover or wildflower honey, but it stands alone with a distinct "waxy" character—a mark of authenticity, not a flaw.

Basswood

Nestled in expansive regions from Southern Canada down to Alabama and Texas, the basswood tree, or American linden, offers up its blossoms to create the remarkable basswood honey. This varietal is not just another sweetener; it's a mead-maker's ally in crafting drinks with character.

Basswood honey is a celebration of clarity; its water-white appearance in its pure form transforms into a beautiful straw color once it becomes mead. Describing its aroma almost applies a vintner's vocabulary—think of the crispness of white wine, the oakiness of chardonnay, and the earthy touch of yeast and minerals.

The flavor is where basswood truly sings. It's like a bite into a green apple, with the tartness and sweetness vying for your attention, followed by a cascade of creamy, buttery, herbal notes, and finishing with a sharp, clean aftertaste. It might be mistaken for its cousin, the Midwestern wildflower honey, but basswood stands alone, with a unique profile.

Another thrilling thing about basswood honey is its adaptability. It's a champion in creating melomels, metheglins, and cysers, embracing and enhancing the flavors of apples and vanilla. It's a honey that encourages exploration—inviting mead-makers to play with its bold palette in their creations.

A word to the wise: while basswood honey brings a distinct and lingering flavor, it can be robust, even slightly intense, when fermented. This quality is not a drawback, but a feature to be harnessed with care, ensuring the resulting mead is balanced and enjoyable.

Blackberry Blossom

As we focus on blackberry blossom honey, let's venture into the verdant land of the Pacific Northwest. Here, the blackberry brambles, teeming with life, share their nectar to create honey as delicate as the first light of day, ranging from a pristine white to the gentle caress of light amber.

In the chalice of mead, this honey pours out a hue like light gold, reminiscent of the dappled sunlight that plays on the forest floor. Bring the glass to your nose, and you're greeted by a floral and subtly leafy aroma, evoking the freshness of the wild outdoors.

This honey's flavour is balanced and distinctive, capturing the understated side of the blackberry's character. It's not the boldness of the fruit that shines through, but the floral, nuanced notes that linger after the blackberries have gone.

Ideal for crafting traditional meads and fruit-forward melomels, blackberry blossom honey is a mead-maker's ally. It marries beautifully with the berries it honors, elevating their taste while maintaining a wonderfully distinct presence.

Blueberry Blossom

Blueberry blossom honey is not what you might first imagine. Sourced from the diminutive white flowers of blueberry shrubs, this honey is the gift of cooler climes—spanning from the Northeastern US to the sprawling woodlands of Michigan and into Canada. It boasts a hue that ranges from the lightest of ambers to a deeper, medium shade.

Blueberry blossom honey provides a deep golden color in a glass of mead, rich and full of promise, yet surprisingly free of any blue or purple tint you might expect from its name. It's a hue that speaks of autumn's golden hour.

Inhaling its aroma, you're met with an array of floral notes, a freshness like green leaves after rain, and a whisper of citrus—complex, yet inviting. Tasting it, you find a harmony of flavors: the fruitiness you'd anticipate, yes, but woven with the verdant notes of foliage and a subtle citrus tang—all culminating in a buttery finish that lingers just long enough, leaving behind a delicate aftertaste.

Standing alone without a substitute, blueberry blossom honey is a distinctive choice for mead-makers. It shines in traditional meads and melomels, where it can subtly enhance the flavors without overpowering them. It's particularly adept at complementing the natural flavors of blueberries, adding depth without dominating with a strong blueberry note.

Buckwheat

Buckwheat honey isn't just any honey—it's a robust, full-flavored cornerstone for the mead connoisseur. Derived from the humble buckwheat plant (which is more herb than grain), this honey has a presence across the upper Midwest of the United States and into Eastern Canada.

This honey's color is a conversation starter: a deep, dark amber that can veer into purple or even black tones. When turned into mead, it transforms the drink into a rich whirl of amber and light brown shades, as if holding onto the Earth's secrets.

Take a whiff, and you're met with an aroma as unmistakable as it is inviting—molasses, treacle, and a hint of sultanas—a story of strength and earthiness. And the flavor? It's nothing short of a bold declaration. Malt syrup, treacle, and caramel cascade into a lingering, earthy aftertaste that's full-bodied and assertive. The taste of tradition and times gone by tells a story.

If you're ever in a spot where buckwheat honey isn't on hand, the richness of avocado can serve as a stand-in, albeit with a different character. This honey is a natural fit for dark braggots, metheglins, and traditional meads, where it can lend its potent flavor profile. It's a perfect match for the warmth of spices like cinnamon and nutmeg, which complement its intensity.

It's worth noting that buckwheat honey is quite the character—it's among the darkest honey and tends to inspire strong opinions. It's a "love-it-or-hate-it" kind of flavor due to its undeniable potency. Its bold nature makes it a honey that plays well with spices. And a little honey trivia for you: the buckwheat honey from the West Coast is a gentler cousin to the robust Midwest variety.

Clover

Let's discuss clover honey, a true classic in the mead-making community. This honey, with its roots in the diverse clover patches from white Dutch to the red blooms found throughout the US and Canada, is a mead-maker's dream. It's everywhere—and for a good reason.

Picture this: the gentle hue of clover honey, a soft range from clear to light amber, much like the color of straw on a sunny day. It brings a similar light amber warmth to meads, giving them an inviting appearance, perfect for any occasion.

And the scent? It's the sweet fragrance of a floral bouquet with a whisper of the clover field—not overwhelming, but just enough to transport you to a sunlit meadow. The flavor is where clover honey shines. It's mild, subtly sweet, the epitome of what many consider "classic honey." It's the kind of taste that brings a comforting familiarity, a sweetness that's just right.

In the world of mead-making, clover honey is often mistaken for wildflower honey due to its gentle, unassuming nature. But that's where its beauty lies—it is a chameleon blending beautifully with other ingredients. Ideal for melomels, traditional meads, metheglins, and pyments, clover honey is like your friend who gets along with *everyone* at the party!

Clover honey has a certain humility to it. It's one of those honeys that doesn't wildly signal for attention. Instead, it offers a mild character that lets the other flavors in your mead do the talking. It's your go-to for a batch that won't be overpowered by more robust flavors, making it a significant player in wildflower honey blends, especially in the Midwest.

Mesquite

Mesquite honey, a true embodiment of the Southwestern US, offers an experience as rich and robust as the mesquite tree from which it hails. In its raw form, this honey gleams from water-white to a sunny amber, casting a straw-to-golden sheen in the meads it graces.

Inhaling the scent of mesquite honey is like a stroll through the desert after rain; it's earthy, with the distinct aroma of raw mesquite wood—earthy, yet distinctly without a hint of smoke. The flavor, consistent with its fragrance, has a depth that lingers on your palate, subtly unfolding with notes of apple or peach, and leaving a full, complex, and distinctively earthy aftertaste that remains on your tastebuds.

If you're looking for a substitute, fireweed honey might come close, but it doesn't match mesquite's bold statement. And while it might remind some of the oak-aged libations due to its woody undertones, mesquite honey is genuinely in a league of its own, offering a rich and characterful flavour profile.

Best suited for traditional meads, metheglins, and melomels, mesquite honey is versatile. It pairs brilliantly with bold flavors like the heat of hot peppers, or the tartness of berries. Think of the kick from chipotle or the sweet, acidic pop of raspberries—mesquite honey can handle them all.

Mesquite honey is noted for its almost intrinsic wood character, a nod to the tree's hardiness. It's one of those rare natural products that bring the essence of their origin into the final taste—imagine the complexity of foods smoked with mesquite wood. Still, here, the honey carries that signature essence—sans smoke.

Orange Blossom

Let's dip into orange blossom honey, a favorite among mead-makers, harvested from the sweet-smelling blossoms of orange trees. These trees spread their roots across the sun-drenched states of California, Florida, Texas, and Arizona. The honey captures the light, ranging in color from the purity of water-white to the palest amber, bringing a golden straw tint to the meads it enriches.

You're enveloped in a heady, floral aroma when you uncap a jar of orange blossom honey! It's bold—reminiscent of a garden in full bloom, laced with the zesty freshness of citrus and the comforting scent of vanilla cream. It's an aroma that's not just sensed with your nose; it's *experienced* and savored.

When it comes to flavor, orange blossom honey doesn't hold back. It's an unforgettable taste, bringing together the subtle sweetness of citrus with a finish that's as smooth as it is lasting. It leaves behind a trace of flowery perfume, a graceful endnote to its vibrant flavor journey.

This honey is a standalone star; it needs no substitutes, no understudies. It's versatile, equally at home in a traditional mead as in the more adventurous melomels. When you pair it with food, it's a harmonious match for the tartness of berries or the gentle heat of spices. It can hold onto its distinctive essence, even when it shares the stage with a robust cast of additions, or the depth of oak ageing.

In mead-making, orange blossom honey is known to be well-rounded, often referred to as a "utility" honey. It's robust and capable of withstanding the test of time and technique. While some may say the Californian variety is superior, boasting a richer, fruitier profile, the Floridian variety also has its charm, offering a subtler, more mellow flavor that some connoisseurs prefer.

Raspberry Blossom

Raspberry blossom honey is a true artisanal delight, hailing from the heart of the raspberry bramble regions—think of the lush Pacific Northwest, the verdant Upper Midwest, and the fruitful Upper East Coast of the US. With its inviting light amber touch, this honey matures into a radiant golden color when in mead, reminiscent of the late summer sun.

Inhale the scent of raspberry blossom honey, and you'll be transported to a sunlit field, surrounded by the gentle buzz of bees and the aroma of citrus-laced floral notes. It's a scent that promises something extraordinary.

The taste? It's a melody of flavors, with the floral notes leading a dance of fruity tangerine, finishing with a delicacy that lingers on the tongue, with hints of the tart freshness of raspberry. This honey brings a mellow smoothness to meads, perfectly balancing the sharpness associated with raspberries, without any bitterness.

Raspberry blossom honey may share some common ground with blackberry honey, but it sings its tune, especially when crafted into traditional meads and melomels. This honey finds its companions in peaches and pears, vanilla, and even chocolate, enhancing those flavors with a natural sweetness that's never overbearing.

Tupelo

In the heart of the Southern swamps, where the white Tupelo trees rise from the wetlands, we find the origins of the exquisite Tupelo honey. Renowned for blossoming in the warmth of Southern Georgia and the Florida panhandle, these trees are spectacular from early April to May.

Gazing at Tupelo honey, you'll be captivated by its light amber glow, a prelude to the golden richness it imparts to mead. This honey comes with an aroma that links to a sun-drenched apple orchard, interlaced with the richness of vanilla and a hint of herbal nuances that bring to mind the greenery of the South.

Savor the flavor, and you'll discover a creamy, sherry-like experience, subtly accented with cinnamon yet surprisingly mild for its full body. It offers distinct and inviting complexity, leaving a smooth, delicate, long-lasting impression.

Tupelo honey is unparalleled in mead-making; no substitutes can mimic its character. It excels in traditional meads and finds harmony in melomels, metheglins, and pyments, where its inherent richness can truly shine. This honey is a perfect companion for the boldness of spices and the robustness of tannic fruits, adding a layer of Southern charm to each blend.

An interesting note about Tupelo honey is its remarkable fructose-to-glucose ratio, which is high enough to ensure it remains fluid and sweet, yet resisting crystallization. This trait makes Tupelo honey one of the sweetest varieties available, and a mead-maker's dream for a consistently smooth mead. When you choose Tupelo honey for your mead, you select a natural sweetness that outshines others, offering a richer and more sumptuous taste.

Wildflower

In the broad art form that is mead-making, each brushstroke of wildflower honey is made up of the wild blossoms of a particular region and season. This honey is a collage, not a single note, made from the nectar of unnumbered blossoms that paint the countryside from one season to the next.

Wildflower honey's color palette is as unpredictable as the wild lands from which it's drawn—ranging from the lightness of early morning mist to the deep amber of a setting sun. When it lends itself to mead, expect a range of flavor that mirrors this natural diversity.

The scent? It's a delicate whisper of the field's bouquet, an olfactory echo of the area's flora. The flavor follows this lead, presenting a floral character as broad as the fields. It's not the boldness of a single flower you'll find here, but the harmonious blend of an entire meadow!

While some might look for substitutes, wildflower honey is irreplaceable for its sheer variability. It may occasionally be mistaken for clover, yet it retains an identity all its own, defined by the complexity and variety of its sources.

As a mead-maker's ally, wildflower honey is exceptionally versatile. It's the perfect partner for meads that boast robust adjunct flavors, from the fruity zest of melomels to the rich depth of braggots. It's a honey that welcomes substantial companions—from the tang of citrus to the richness of spices—allowing them to shine.

Commercially Blended Honey

Commercially blended honey offers a dependable base ingredient for those in the mead-making community. It's a go-to choice for its consistency and standardization.

Let's unpack this: much of the commercially blended honey undergoes a heat pas-teurization process, which occurs at about 145°F (63°C). It's a step taken to ensure the honey's cleanliness, but it's worth mentioning that this might slightly alter some of its delicate natural flavors.

The real charm of commercially blended honey lies in its balanced pH, typically around 3.9, which provides a stable foundation for fermentation. Its light amber color reflects a composition that's not so heavy on minerals, ensuring a clean, clear mead without unwanted flavors that might come from a higher mineral content.

What stands out about this type of honey is its neutral flavor profile. It's intentionally crafted to be unobtrusive in taste and aroma, making it an excellent starting point for mead-makers. This neutrality allows the other ingredients in a mead recipe to take center-stage, letting you craft a beverage highlighting your unique additions.

In terms of quality, commercially blended honey often comes with a guarantee of low moisture content. This is more than just a technical detail—it's crucial for avoiding fermentation problems and achieving a high-quality end product. The consistency in color also simplifies recordkeeping, which is a boon for those who appreciate meticulous documentation of their mead batches.

Choosing the Ideal Honey for Your Mead

Selecting the right honey for your planned mead requires a connoisseur's touch. It's much like a chef choosing the perfect spice, or a painter choosing the right shade. The honey you pick will lay the foundation of your mead's flavor profile—whether you're after the boldness reminiscent of ancient brews, or a subtle sweetness that teases the palate.

Consider the flavor you want to shine in your mead. With the myriad of honey varieties, each brings its character to the brew. Some offer a strong, almost earthy intensity; while others are milder, with a delicate floral presence. The key is to find a honey that will complement and enhance your other selected ingredients, creating a harmonious blend in the finished mead.

Each variety of honey reflects its environment, imbued with unique characteristics that will significantly influence the flavor of your mead. Look into the origins, the stories, and the flavor notes of various kinds of honey. This exploration not only educates, but also inspires, helping you make an informed choice that's tailored to your mead vision.

Before you commit to a honey—taste it. If that's not possible, seek out descriptions or, better yet, recommendations from fellow mead-makers. Tasting the honey on a neutral base, like bread or crackers, is like sampling a fine wine—it's about the pure experience of flavor and aroma.

Lastly, let practicality be your guide, as well. Your budget and the honey's availability are factors that can't be ignored. While exotic honeys are tempting, consider their cost and the volume you need for your mead. Rarity and price can vary, so weigh these against the scale of your mead-making endeavor.

Where to Source Your Honey

Local is Lovable

When you choose local honey for your mead, you're not just buying a product; you're experiencing a living story. Each local batch carries the essence of its landscape, a reflection of the flora and fauna of the area. This local touch adds a unique dimension to your mead, infusing it with the character of your region.

Moreover, buying local honey fosters a connection with your community. By engaging with local beekeepers, you gain insights into the honey's origin and the bees' well-being. These relationships often lead to discovering rare, single-origin honey that can lend extraordinary flavors to your mead.

However, remember that local honey's availability and price can fluctuate. This scarcity often adds value to the honey, making it a prized ingredient for particular batches of mead. The varied nature of local honey also encourages experimentation, inspiring mead-makers to adapt and create diverse flavor profiles.

Big Store Bargains

Turning to large retailers like Costco or Sam's Club for honey is a savvy and economical choice for mead-makers, especially when balancing quality with budget. These stores offer a variety of honey in quantities that cater to both small-scale trials and larger mead-making ventures. Standard varieties such as clover and wildflower honey are found here—perfect for those just beginning their mead-making practice, providing a harmonious flavor base suitable for a range of recipes.

The beauty of these larger stores lies in their consistent honey supply. This reliability is a boon for mead-makers who seek a steady source of ingredient, free from the worries of seasonal shortages or erratic availability. While this honey is budget-friendly and reliable, it may lack the unique regional flavors that local honey offers. This factor alone might be pivotal for mead artisans on a quest to encapsulate the distinctive tastes of their local environment in their brews.

The Online Honeycomb

The digital marketplace has transformed how we source honey for mead-making. Take Amazon, for instance, with its vast array of honey options that appeal to various tastes and preferences. You're not limited to common varieties; exotic ones like orange blossom or buckwheat honey are just a click away, each bringing a unique edge to your mead's flavor.

What's more, the ease of online shopping lets you explore honey from across the globe—right from your home. This opens up a world of flavors and encourages adventurous blending. And for those planning more significant mead projects, buying honey in bulk online can be a cost-effective move. Remember, it's crucial to pick reputable sellers to ensure you get quality honey, as the online market is vast and varied regarding reliability and authenticity.

Bulk Buying Brilliance

Venturing into bulk buying marks an exciting phase in mead-making. Suppliers like Dutch Gold Honey, offering vast quantities like 60-pound pails, provide a cost-efficient option for those scaling up their production. This choice, however, also means committing to a single honey variety over multiple batches. While this can deepen your understanding of a specific honey type's flavor nuances, it narrows the range of flavors to experiment with. Bulk buying balances economic efficiency with consistent flavor experiences in larger-scale mead production, and can be ideal for those who prefer a specific honey.

A Note on Quality

To pursue quality honey for mead-making, focus on raw, unpasteurized varieties. They're rich in natural flavors and characteristics, essential for crafting a mead with depth and complexity. While pasteurized honey is an option, it might not offer the same intricate flavors.

Starting Small, Dreaming Big

Are you beginning your mead-making adventure? Yes? Then begin with smaller honey quantities. This strategy lets you play around and understand the nuances of different honey types without the pressure of large-scale production. As you grow in confidence and skill, gradually increase your batch sizes. This step-by-step journey hones your craft and guides you toward creating mead that genuinely resonates with your vision, taste and audience.

Water Choices for Mead-Making

Water Bottle/Credit: INSAGO (www.Shutter stock.com)

Choosing the right water for mead-making starts right from your tap. It's a great candidate for your mead if the water tastes good. However, steer clear of distilled or deionized water, as these lack the beneficial minerals your mead needs. Hard water, rich in carbonates, is often the hero here. Choosing hard water for mead-making is particularly advantageous. This is because the carbonates in hard water act as a natural buffer. They prevent the pH levels from becoming too acidic during fermentation—which is crucial. If the pH drops too low, it can hinder the yeast's ability to complete fermentation efficiently.

It acts as a buffer, keeping the pH levels in check during fermentation and ensuring that the yeast thrives without getting overwhelmed by acidity.

Be mindful of the water treatments used in your area, particularly chlorine and chloramine. These can lead to unwanted flavors in your mead, so finding a way to remove them, or choosing an alternative water source is very important.

While boiling can quickly remove chlorine; chloramine requires a different approach. Adding a Campden tablet containing potassium or sodium metabisulfite to your water can effectively cleanse it of chlorine and chloramine.

A Campden tablet is a small, compressed tablet primarily composed of sodium or potassium metabisulfite. It is used in brewing and winemaking for sterilizing and stabilizing purposes. Its primary function in mead-making is removing chlorine and chloramine from water, ensuring a pure base for fermentation. Campden tablets are readily available at home brewing and winemaking supply stores (in physical and online locations).

Allow the water to rest after treatment to ensure a complete reaction. Additionally, consider using an activated charcoal filter. An activated charcoal filter, used in water purification, employs activated charcoal to absorb impurities and remove contaminants like chlorine, chloramine, and organic compounds. It won't alter the water's hardness but will remove any unpleasant odors or flavors, contributing to a purer base for your mead.

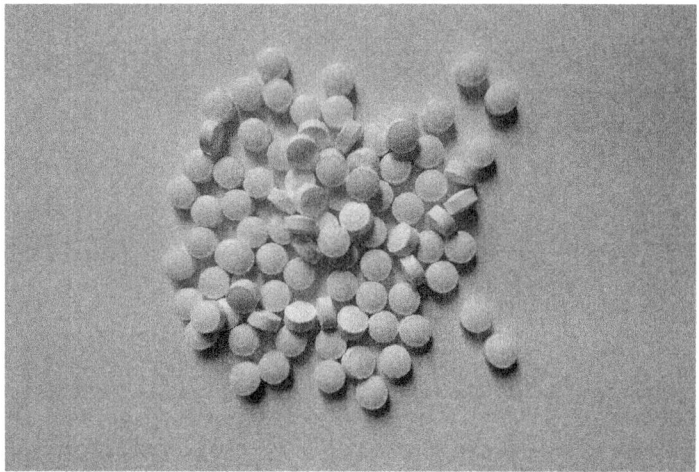

Campden Tablets/Credit: Apugach (www.Shutterstock.com)

Using Bottled Water in Mead

Turning to bottled water for mead-making is smart, particularly for those who doubt the quality of tap water. Natural spring water stands out as a top choice among the various types of bottled water available. Its inherent mineral content can subtly but significantly enhance the flavor of your mead, adding a layer of complexity to the final product. When selecting bottled water, avoid brands that add minerals or flavors, as these additives can (unpredictably) alter the taste of your precious mead.

Bottled water ensures a consistent, controlled quality, a critical factor in achieving a reliable outcome in your mead-making. Generally, the purity of bottled water means there's no need for boiling, as it's typically devoid of common tap water contaminants, like chlorine or chloramine. However, boiling becomes a viable option when the source or purity of bottled water is questioned. This step can provide peace of mind, ensuring that your water base is as pure and neutral as possible, allowing the nuanced flavors of your honey and other ingredients to shine through in your mead.

Yeast

Dry Yeast/Credit: Gummy Bear (www.Shutterstock.com)

As you are starting on your mead-making journey, you'll soon find the choice of yeast isn't just a minor detail; it's a cornerstone of your craft. *Saccharomyces cerevisiae,* a species renowned in wine and beer-making, is equally pivotal in mead. This species is a treasure trove of diversity, offering a myriad of strains, each with its unique effect in the fermentation process.

Fermentation: A Prelude to Yeast Selection

Starting with a clear understanding of fermentation is pivotal to your mead-making practice. This isn't just a process; it's the very heartbeat of your mead! Learning its nuances aids tremendously in choosing the yeast with which to bring your mead to life.

Picture fermentation as a sophisticated piece of choreography where yeast, water, sugar, and nutrients all play their parts in perfect harmony. In this controlled environment, yeast thrives, converting sugars into alcohol and carbon dioxide. The choice of honey and the intricacies of your recipe set the stage for this rather enchanting process. It's like conducting an orchestra—every element must come together, just right, in tone and timing.

But let's not forget, this stage is also a tightrope walk. Any imbalance can lead to those dreaded off-flavors. Imagine the yeast as diligent workers in your meadery. If they're unhappy and lack oxygen nutrients, or the pH isn't right, they protest by producing unwelcome flavors and aromas. Our collective quest is to achieve a clean fermentation—free from these pitfalls.

The process of fermentation is divided into four phases:

1. **The Introduction**—The Lag Phase: This is where yeast first meets the must, a crucial time for them to get acclimatized.

2. **The Growth Spurt**—Respiratory Phase: Oxygen plays a starring role here, helping yeast grow and multiply.

3. **The Main Event**—Actual Fermentation Phase: This is where the magic happens, as yeast transforms sugars into the soul of your mead.

4. **The Curtain Call**—Flocculation: The finale, where yeast takes a bow and settles down, signaling the end of the fermentation performance.

It's in your hands to create the ideal conditions for each phase. Keep an eye on the little things—the yeast's nutritional needs, the temperature, and the oxygen balance. When everything aligns, the fermentation process, typically wrapping up in about two weeks, leaves you with a mead that's nothing short of perfection.

We will now roll up our sleeves and explore each stage in greater detail! Here we go!

Step 1: The Lag Period

Imagine the lag period in fermentation as the time for the yeast to acclimate to its new environment, much like a traveler arriving in a foreign land. The yeast must have just the right sugar blend and a welcoming pH level in this phase. Here, the yeast starts its crucial preparations for the passage ahead.

In this initial stage, the yeast cells are in a state of bustling activity, similar to a team of athletes gearing up before a big game. They absorb nutrients, fortify their cell walls, and prepare for an intense period of growth and reproduction. It's a time of building up reserves and strengthening themselves to create healthy new cells.

Step 2: The Respiratory Phase

Next comes the respiratory phase, a vital period of growth. Here, the initial population of yeast begins a rapid multiplication process. It's a fascinating display of nature's efficiency, where each yeast cell forms a bud—a clone of itself—which eventually breaks away as an independent, new cell.

This process is like a masterful biological symphony! The yeast cells don't just split; they undergo a complex budding process. Each cell develops a bulge, receives a copy of the yeast's genetic material, and eventually separates into a new yeast cell. This phase is a remarkable feat of natural engineering, requiring significant resources, especially in crafting fresh, robust cell walls.

The creation of these cell walls is a critical aspect of this phase. They must be sturdy enough to protect the cell, yet permeable enough to allow for the essential sugar metabolism process. This delicate balance requires a blend of oxygen, nitrogen, and various nutrients, illustrating the intricate relationship between the yeast and its environment.

This stage is often called "aerobic" or "respiratory" for a good reason. The yeast's need for oxygen during this time is paramount. Different yeast strains might have varying demands for oxygen and nutrients, but they all rely on these elements to some extent. This phase is necessary for setting the stage for the primary fermentation event, highlighting the importance of creating the *right* conditions for your yeast to thrive.

Step 3: The Heartbeat of Fermentation

As we move into the actual fermentation phase, it's like entering the center of a bustling city—where the real action happens. This phase marks a significant shift from the oxygen-rich respiratory stage, to an anaerobic environment where the focus is no longer on growth—but on transformation.

In this third stage, the yeast, now acclimated and mature, turns its attention to the sugars in the must. It's a meticulous process, like an artist turning a blank canvas into a masterpiece! The yeast absorbs the sugars, converting them through complex bio-chemical reactions into alcohol and carbon dioxide. This phase really is the essence of mead-making, where the sweet must begin its journey towards becoming the mead we so cherish.

The pace of this transformation varies. Depending on your choice of yeast and the nutrient composition of your must, fermentation can begin within 12 to 72 hours after introducing the yeast. Generally, the process starts in about a day, with a well-pre-pared must and adequate yeast. During this time, you'll notice the yeast population hitting its peak and then gradually declining—a natural ebb and flow in response to the changing environment within the fermenter.

Step 4: Flocculation

Flocculation is the grand finale of the fermentation process! As the sugars deplete and fermentation slows, the yeast cells remarkably transform. They begin to clump together, forming what we call "flocs." This phase is like playing the final bars of a symphony, where everything comes together in harmony.

The characteristics of this stage can vary dramatically between different yeast strains. Some settle quickly and compactly at the bottom of the fermenter, creating a clear, refined mead. This ability to settle well and form a dense layer is a trait highly valued in yeast strains, as it directly influences the clarity and quality of the final mead product.

Selecting the Right Yeast for Your Mead

Choosing the right yeast is one of your most crucial decisions. While honey, water, and other ingredients lay the foundation of your mead, the yeast diligently works to weave these elements into the delightful beverage that mead enthusiasts adore. The selection of yeast is not just a choice; it's a skilful art that considers various factors, including the unique ingredients of your mead, the alcohol by volume (ABV) you aim to achieve, and the temperature conditions under which your mead will ferment. It's important to understand that different yeasts have unique reactions to these factors, making your decision critical to the outcome of your mead.

Gravity in Mead-Making

hydrometer measuring specific gravity/Credit: BDoss928 (www.Sh utterstock.com)

Before we explore yeast selection, grasping the concept of "gravity" in mead-making is important. Gravity measures the concentration of dissolved solids—primarily sugars—in your must, or mead. The higher the concentration, the higher the gravity. In mead-making, this plays a pivotal role in yeast selection and fermentation.

There are several ways to measure gravity, each offering insights into the density and sugar content of your must:

- **Specific Gravity:** Commonly used in home brewing, this scale compares the density of your must to water. For example, a specific gravity of 1.050 indicates that your must is 5% denser than water.

- **Degrees Plato/Balling:** These scales are more prevalent in professional brewing circles. They quantify the percentage of sugar by weight in your must. For instance, a reading of 12° Plato means 12% of the must's weight is sugar.

- **Brix:** This scale is a staple in winemaking and measures sugar content in solutions, much like Plato, but is typically applied to fruit juices and wines.

A calibrated hydrometer or refractometry can measure gravity, especially in more advanced brewing setups. In the next chapters, I'll walk you through how to utilize these tools effectively.

The standard reference point for all these measurement systems is pure water at 68°F (20°C), with a gravity of 1.000. For example, a must composed of 1 gallon of honey and enough water to make up 5 gallons may have a specific gravity of around 1.094 or 22.5° P.

One critical factor in yeast selection is the strain's tolerance to gravity. Some yeast strains may struggle to initiate fermentation in high-gravity must, potentially reverting to a dormant state. Choosing a yeast strain with a robust gravity tolerance is key if your mead recipe calls for a high initial gravity. As the fermentation process gets underway, the sugars, which contribute to the must's gravity, convert into alcohol and CO_2, consequently lowering the gravity. Ethanol being lighter than water can result in a fully fermented mead having a specific gravity below 1.000.

Measuring the specific gravity before and after fermentation is not just a routine check; it's an insight into the potential alcohol content of your mead.

In the chapters ahead, I will guide you through calculating the original gravity of your mead. This knowledge is crucial in selecting the yeast for your must's gravity. Though it may sound initially complicated, we'll break it down together, so you are fully equipped to make informed decisions in your mead-making practice.

Alcohol Tolerance of Yeast

Every yeast strain has an "alcohol tolerance" rating, indicating the maximum alcohol percentage it can feasibly ferment before halting. However, the world of yeast is full of surprises! Often, they exceed these specified limits, creating higher alcohol levels than anticipated. Yeast typically ceases fermentation when it exhausts its sugar supply or is overwhelmed by the alcohol it produces. While the alcohol tolerance rating is helpful—expect some exciting variations.

Selecting the Perfect Yeast for Your Brew

Now that we've laid the groundwork by exploring the essentials of fermentation and the concepts of gravity and alcohol tolerance, let's take an exciting step forward into the world of... yeast selection.

When hunting for the ideal yeast for your mead, think of it as finding a dance partner who knows the steps and brings their flair to their performance. Here's what to keep an eye out for:

- **Flavor and Aroma:** Every yeast strain is like a character in a story, each bringing its unique flavor and aroma to the tale of your mead. This choice is foundational, as it sets the stage for the sensations your mead will offer you. It's like selecting the key ingredient in a gourmet recipe—crucial and transformative.

- **Alcohol Tolerance:** This is where you match the yeast's strength with your vision for the mead. High alcohol tolerance yeasts are like marathon runners, going the distance for stronger meads; while others are sprinters, perfect for a lighter, more delicate brew.

- **Temperature Range:** Yeasts have their preferred climates, (just like us). Some thrive in the warmth of a summer day, others in the coolness of a spring morning. Selecting a yeast that feels at home in your brewing environment is key to a smooth and successful fermentation.

- **Nutrient Requirements:** Mead's base, honey, is naturally deficient in some nutrients that yeasts crave. Understanding what your yeast needs in order to thrive is like knowing the perfect soil mix for a rare plant. Some yeasts are high-maintenance, requiring a rich blend of nutrients, while others are more low-key.

- **Flocculation:** This trait is about how the yeast settles once the party ends. High flocculating yeasts clear out neatly, leaving behind a clear, stable mead. Low flocculators might need a bit of coaxing to settle down. This characteristic impacts the clarity and visual appeal of your final mead.

The following chapters shed more light on these characteristics. Understanding them in more detail will enhance your knowledge and refine your skill in crafting meads that resonate with your style and taste. You are now well on your way to knowing more about mead-making and fermentation!

Choosing Between Dry and Liquid Yeast

Choosing between dry and liquid yeast is pivotal, bringing unique effects to fermentation. Dry yeast, celebrated for its resilience and longevity, is sealed in foil packets that guard against humidity, ensuring a stable environment for the yeast cells and their initial nutrient boost. Ideal for long-term storage, its vitality does wane over time, with refrigeration recommended to prolong its life.

Liquid yeast, on the other hand, is a beacon for diversity, available in smack-packs and vials, offering an array of strains to fine-tune the flavor profile of your mead. With their innovative design, smack-packs include a nutrient sack to invigorate the yeast. At the same time, vials are densely packed with yeast, ready for direct introduction to your brew under specific conditions.

Both forms require careful handling, being stored cold, and being brought to room temperature before their moment to shine.

Mastering Yeast Preparation

It's essential to recognize that how you prepare your yeast, whether dry or liquid, lays the foundation for a successful fermentation. Even though dry and liquid yeasts often come with a "ready to use" label, giving them a little extra care before introducing them to your must can make a difference. For dry yeast, this means embracing the art of proper rehydration. Creating a starter could be your key to success if you're working with liquid yeast. We'll explore these techniques, each important in its own right, to ensure your lovely mead reaches its full potential.

Preparing a Starter for Liquid Yeast in a Vial

Liquid yeast in a vial

49

Creating a yeast starter is a critical step in the mead-making process, ensuring your liquid yeast culture flourishes and leads to a successful fermentation. When you opt for liquid yeast, boosting the yeast cell count before introducing it to your must can significantly impact the vigor and health of your fermentation. A well-prepared starter fosters a thriving yeast population, which is critical for a robust start. It's about setting the stage for those tiny yeast cells to perform at their best, aiming for a bustling community of around 1 million cells per milliliter in your must. While counting cells might be beyond the scope of most home mead-makers(!) employing a starter is a tried and true method to guarantee your yeast culture is primed for action, ready to transform your must into delightful mead.

The goal is to craft a starter that mirrors about 1% of your must's total volume, focusing on the yeast slurry's density. A standard 5-gallon batch equates to roughly 6 to 7 ounces (about 0.2 liters) of slurry. Achieving this requires a starter volume of around 10% of your batch size or roughly half a gallon, though some might suggest a gallon. Experience has shown that a 10% volume is adequate, negating the need for larger quantities. The starter solution should closely resemble your must in composition, with an ideal gravity of approximately 1.050. A nutrient-rich, lower-gravity starter encourages the development of vigorous, healthy yeast cells ready to tackle the task of fermentation.

Once you have selected your yeast, your next steps include preparing a clean, sanitized 2-liter container, and sterilized brewing essentials—like a funnel, airlock, and bottle cap. The starter solution is simple yet vital, consisting of water, yeast energizer and nutrient, plus honey, boiled together and then cooled to room temperature in the 2-liter container you previously prepared. Introducing the yeast to this hospitable environment, followed by vigorous shaking or oxygenation, sets the stage for fermentation to kick off within 12 to 24 hours.

As you're ready to add your yeast to the must, gently remove most of the liquid no longer needed, leaving behind the yeast mixture. Give the mixture a good stir to ensure it's fluid enough for pouring, then transfer it into your must that's waiting. Distilled or preboiled water is beneficial for washing residual yeast from the container, allowing you to incorporate every last bit into the must. Fermentation begins relatively quickly after your must is ready and you've introduced your selected yeast.

How to Prepare a Smack Pack

Liquid yeast in a smack pack

Incorporating a smack pack into your mead-making is a fascinating and crucial step to ensure your mead ferments to perfection. Begin by acquainting yourself with the pack, gently feeling around to locate the inner nutrient sack. Once located, coax the nutrient sack towards one end of the pack—this paves the way for activation. With the pack positioned firmly on a table, a firm, confident press with your palm is required to break the inner sack. (This might test your resolve, but trust the pack's design to withstand the pressure.) After the small victory of breaking the sack, give the pack a vigorous shake, mixing the nutrients with the yeast—it´s a bit like awakening a sleeping giant!

Patience now plays its part. The pack needs a warm spot to rest, like letting the dough rise, swelling to about 1.5 to 2 inches thick. This transformation is not instantaneous; depending on the pack's age, it might take several days for the yeast to be primed for its pivotal role. Remember, time itself is an ingredient in making mead, as much as honey or water.

As the pack nears readiness, a day added for each month past its creation ensures its potency. Before introducing this primed yeast to the must, a sanitation ritual is paramount, ensuring no unwanted guests join your mead-making team. Gradually acclimatizing the yeast to the must's temperature is the final act of care before the magic happens, supporting a smooth transition for the yeast into its new environment.

Dry Yeast Rehydration

flask with rehydrated yeast/Credit: rossiaa33 (www.Shutterstock.co m)

In the past, the common practice was to pitch yeast directly into the must. This method was straightforward—sprinkle the dry yeast on the surface, wait 15 to 30 minutes, and then stir. However, as we've come to understand yeast's complexity, this approach has revealed some shortcomings.

The contemporary mead-maker looks beyond just awakening the yeast. We now understand the importance of rehydrating yeast, but with a twist—adding micronutrients. Imagine prepping a marathon runner with the proper nutrition; similarly, these micronutrients equip the yeast to face the stresses of the fermentation process with robustness.

Directly pitching yeast into the must, once standard practice, is now considered less than ideal. Here's why:

- **Clumping Issues:** When you sprinkle dry yeast directly into the must, it clumps. This clumping can prevent the yeast from rehydrating evenly, affecting its performance.

- **Vulnerability to Must Components:** They must harbor substances like sulfur dioxide or residual fungicides. These can be lethal to dry yeast during rehydration. Once the yeast is rehydrated, it can handle these elements to some extent, but not during the critical water uptake phase.

- **Risk of Off-Flavors:** Under the stress of direct pitching, yeast cells can leak substances that may lead to undesirable flavors in your mead. It's like putting the yeast under unnecessary pressure, and the results are often unpleasant.

- **Compromised Yeast Viability:** Pitching yeast directly into the must can reduce its viability, leading to weaker fermentation. More yeast *isn't* a solution, as the risk of off-flavors still looms.

Unlike the more protracted process of preparing a liquid yeast starter, rehydrating dry yeast is quick and straightforward. Dry yeast packets are affordable and convenient, making them popular among mead-makers. The rehydration process only takes a few minutes, contrasting to the day or two required for a starter. This efficiency is one of the many reasons why dry yeast is so appealing.

Dry yeast has an excellent shelf life, another feature that adds to its appeal. Some suppliers even suggest keeping unopened packets in the freezer for extended freshness. But remember, bringing the yeast back to room temperature before you start rehydrating is essential. This step ensures the yeast wakes up correctly, and is ready for action.

The first few minutes of rehydration are a critical transformative phase for the yeast. It's like waking up from a deep sleep and immediately running a race—the yeast must adapt quickly from dry to active form. In this stage, the yeast cells soak essential minerals from the rehydration liquid. It's crucial to avoid distilled or deionized water here; without minerals, the yeast could be harmed—even destroyed.

Temperature is a critical factor in this process. It's too cool, and you risk a significant loss in yeast viability. For instance, rehydrating in water at 60°F can slash viability by 60%. Most yeast suppliers recommend a temperature of around 104°F for optimal results.

Using a product like Lallemand Go-Ferm during rehydration can make a difference. This unique blend of micronutrients, (including vitamins, minerals, and amino acids), is tailor-made to prepare the yeast for its fermentation journey.

Here's how you do it: start with a sanitized glass container, like a measuring cup. As a rule of thumb, for every 4 grams of dry yeast, you'll need 5 grams of Go-Ferm and about 4 ounces of water (later in this book, I will introduce you to a more precise method to estimate your exact needs of Go-Ferm and water). The water temperature should be around 104°F, as recommended by the yeast vendor. Dissolve the Go-Ferm in the water, add the yeast, and stir gently. Let it sit for 15–30 minutes; going beyond 30 minutes could reduce the yeast's effectiveness.

During rehydration, cover the container with plastic cling wrap. This simple step safeguards against wild yeast and bacteria, ensuring that only your *chosen* yeast works on your mead.

Yeasts for Beginner Mead-Makers

I'm thrilled to now guide you through a handpicked selection of yeasts, ideal for those venturing into the world of mead-making. In this section, you'll find a comprehensive overview of each yeast variety, tailored to assist beginners in navigating their choices. I'll be covering the specific attributes of each yeast, pinpointing which meads they best complement, and giving you a glimpse of the unique tasting notes they can bring to your beverage creation.

I encourage you to further explore the yeast that piques your interest. There's a wealth of information available online that can offer further insights into the optimal use of your chosen yeast. Additionally, I highly recommend downloading the datasheet for the yeast from the producer's website. A quick Google search should lead you to these datasheets, packed with all the essential details you need to use the yeast successfully. This step is more than just a recommendation—it's key to ensuring your mead-making journey is as informed and rewarding as possible. Here is the low-down on some good, relevant yeasts:

Safale S-04

Originating from England, Safale S-04 is an ale yeast celebrated for its rapid fermentation ability. It's not just about speed; this yeast brings a delightful harmony of fruity and floral notes, adding a nuanced depth to your mead.

While traditionally a go-to for American and English Ales, the versatility of Safale S-04 makes it an excellent match for various mead styles. It shines in bright fruit melomels, where its fruity essence complements the natural fruit flavors. Traditional meads and braggots also benefit from its floral notes, enhancing their complexity and appeal.

Specifications:

- **Ester Production:** Low total esters from Safale S-04 mean your mead retains a cleaner, purer profile, letting the ingredients' natural tastes take center stage.

- **Attenuation Power:** With an attenuation range of 74–82%, this yeast balances the sweetness and alcohol content, giving you control over the final taste.

- **Sedimentation:** Its fast sedimentation quality is a boon, aiding in quicker clarification and a smoother mead.

- **Alcohol Tolerance:** Safale S-04 can comfortably handle an alcohol content of 9–11%, catering well to a range of mead styles without overwhelming them.

- **Fermentation Temperatures:** The yeast thrives between 18–26°C (64.4-78.8 °F), offering flexibility to adapt to various brewing conditions.

- **Nitrogen Needs:** Its medium nitrogen requirement should be noted, ensuring you prepare your must accordingly for the best results.

Safale US-05

Safale US-05 is an American ale yeast renowned for its versatility and exceptional qualities, making it a superb choice for various mead styles. Famous for its role in American ale production, this yeast is also known for creating neutral and well-balanced meads, marked by a distinct clean and crisp finish. Another notable feature of Safale US-05 is its ability to form a firm foam head, adding an aesthetic appeal to certain mead styles. Moreover, its excellent suspension quality during fermentation ensures a consistent and reliable brewing process.

While its roots are in brewing beer, Safale US-05's adaptability makes it also a stellar performer in mead-making. It's particularly effective in traditional meads, whose neutral profile allows the honey's natural flavors to shine. This yeast is also an excellent choice for hopped meads and various melomels, where it complements the ingredients without overpowering them. Its balanced fermentation ensures that the mead retains the authentic character of its components—even after fermentation.

Specifications:

- **Fermentation Range:** Thriving in a temperature range of 64.4–78.8°F (18–26°C), Safale US-05 offers flexibility, making it suitable for various brewing conditions.

- **Attenuation Level:** With an attenuation of 78–82%, it efficiently converts sugars into alcohol, achieving a delightful balance in sweetness.

- **Alcohol Tolerance:** Its alcohol tolerance of 9–11% makes it ideal for meads with moderate alcohol content.

- **Nitrogen Needs:** The medium nitrogen requirement of this yeast should be factored in during must preparation to ensure the yeast performs at its best.

Safcider

Among the myriad choices, Safcider Yeast is an intriguing option, especially for those exploring fruit-forward meads.

What sets Safcider Yeast apart is its delicate yet complex aromatic profile. It's a symphony of fresh apple notes blended with the richness of applesauce-like undertones. This harmonious combination results in a mead that boasts a balanced mouthfeel, paying homage to the classic structure of ciders. Initially observed in French cider recipe trials, this attribute of the yeast opens up possibilities for mead-makers seeking to infuse their brews with a nuanced fruit character.

Safcider Yeast isn't confined to cider-making; its versatility also shines in the mead-making arena. It's an excellent match for cysers, where the interplay of apple juice or cider with honey creates a delicious concoction. Meads incorporating malic fruits also benefit from this yeast's ability to enhance and respect their natural flavors. Its proficiency in handling fresh and concentrated apple juices, and navigating challenging fermentation conditions like sugar syrups, make it a resourceful ally in your mead-making adventures.

Specifications:

- **Killer Factor:** Being sensitive, Safcider Yeast requires a bit of mindfulness in mixed fermentation settings.

- **Temperature Flexibility:** With a broad fermentation temperature range from 50–86°F (10–30°C), it offers the flexibility to brew in varying conditions, suiting both cool and warm environments.

- **Nutritional Needs:** Its low nitrogen requirement simplifies your fermentation process, making it a hassle-free option for beginners and seasoned brewers.

- **Alcohol Tolerance:** The high alcohol tolerance of up to 18% v/v is a notable feature, catering to stronger meads that pack a punch in alcohol content.

Lalvin EC-1118

Lalvin EC-1118, a yeast known for its robustness and versatility, is a standout choice for mead-makers. Lalvin EC-1118 hails from a prestigious sparkling wine region, bringing a legacy of producing excellent base wines and aiding in secondary bottle fermentation. Its standout feature is its resistance to osmotic pressure, making it a reliable choice for mead-makers. This yeast is celebrated for its vigorous kinetics, a consistent and robust player in the fermentation process.

What sets Lalvin EC-1118 apart is its sensory neutrality. It adds minimal flavor or aroma to the mead, allowing the natural characteristics of your other ingredients to shine. This trait has made it a popular choice for producing white and red wines globally, and holds similar promise in mead-making.

Lalvin EC-1118's neutral profile makes it ideal for meads where the yeast's character should be subtle, allowing the flavors and aromas of the mead's base ingredients to take center stage. Due to its energetic fermentation nature, it's particularly well-suited for flavor-rich brews, as it can sometimes overshadow softer aromas. This yeast is a perfect match for high-alcohol meads thanks to its exceptional alcohol tolerance.

Specifications:

- **Aroma:** Neutral, ensuring that the intrinsic aromas of your mead ingredients are not overpowered.

- **Alcohol Tolerance:** Capable of handling up to 18% alcohol, Lalvin EC-1118 is ideal for crafting potent meads.

- **Fermentation Temperature Range:** With a broad range of 50–86°F (10–30°C), this yeast is adaptable to various brewing conditions.

- **Nitrogen Needs:** Its low nitrogen requirement reduces the complexity of nutrient management in your mead-making process.

Lalvin D47

Lalvin D47, a yeast with unique attributes, is an excellent choice for creating meads with depth and complexity. Initially isolated in France's Côtes du Rhône region, it was selected from a pool of 450 isolates for its outstanding fermentation properties. Known for its role in malolactic fermentations, Lalvin D47 is especially suited for meads requiring a substantial mouthfeel, such as those with higher ABV or rich in dark berry fruits.

What's truly remarkable about Lalvin D47 is its ability to release polysaccharides into the must during fermentation. This process lends a round, soft palate to the mead, with a substantial weight that's very desirable in certain styles.

Lalvin D47 shines in meads where a full-bodied and aromatic profile is planned. It's a perfect match for barrel fermentation, where its attributes can be fully expressed. This yeast excels in meads featuring tropical or citrus fruits and is a fantastic choice for meads with a bigger mouthfeel. The soft aromas and vigorous fermentation make it an ideal companion for flavor-rich brews.

Specifications:

- **Fermentation Range:** With a range of 59–86°F (15–30°C), Lalvin D47 offers versatility in various brewing conditions.

- **Alcohol Tolerance:** It can handle up to 15% alcohol, catering to a range of mead styles.

- **Nitrogen Requirement:** The low nitrogen need of this yeast simplifies the fermentation process, making it a practical choice.

- **Killer Factor:** Possessing a yeast killer factor, Lalvin D47 can be an essential player in mixed fermentation settings.

- **Compatibility with Malolactic Bacteria:** It has a very high compatibility with malolactic bacteria, enhancing its suitability for complex fermentation processes.

Lalvin 71B-1122

Among the diverse yeast options, Lalvin 71B-1122 emerges as a particularly appealing choice for certain styles of mead, thanks to its unique fermentation qualities. Renowned for its capacity to absorb polyphenolic compounds, this yeast can subtly influence the tannin structure in meads made with dark berries and red fruits. This ability to manage tannins is a boon for mead-makers aiming to craft a beverage with a smoother, more rounded profile.

Lalvin 71B-1122 is a favorite among mead-makers, particularly for brews that integrate dark berry flavors or red fruits. Its unique properties make it ideal for meads where a smoother taste and a fruit-forward profile are desired. This yeast can bring out the best in your ingredients, allowing their natural flavors to shine.

Specifications:

- **Fermentation Range:** Operating within 59–86°F (15–30°C), Lalvin 71B-1122 offers versatility in fermentation conditions.

- **Alcohol Tolerance:** With an alcohol tolerance of up to 14%, it's suitable for various mead styles.

- **Competitive Killer Factor:** This sensitive strain requires careful consideration in mixed fermentation environments.

- **Malic Acid Consumption:** High malic acid consumption is vital to producing smoother, fruitier meads.

- **Nitrogen Needs:** The low nitrogen requirement of Lalvin 71B-1122 makes the fermentation process more manageable.

Lalvin QA23

Lalvin QA23 is a unique and versatile yeast, particularly adept at enhancing specific fruit profiles in meads. Originating from the esteemed Vinhos Verdes region in Portugal, this yeast is celebrated for bringing out citrus fruit aromas, such as lime and grapefruit, adding a zesty tone to your mead.

Lalvin QA23 is renowned for its clean and efficient fermentation process, ensuring the yeast does not overshadow your mead's natural fruit flavors.

Lalvin QA23 is particularly notable for its high β-glucosidase activity. This enzyme plays a crucial role in transforming non-volatile aromatic compounds into more aromatic volatile forms, thus contributing significantly to the varietal expression of the fruit used. While it's a popular choice for enhancing the flavor profile of Sauvignon Blanc, its characteristics also make it ideal for tropical-fruited and citrus-fruited meads. However, it's worth noting that some mead-makers have found it less suitable for traditional meads.

Specifications:

- **Fermentation Range:** This yeast operates within a temperature range of 57–82°F (14–28°C), offering versatility for different brewing environments.

- **Alcohol Tolerance:** With an alcohol tolerance of up to 16%, it's suitable for a variety of mead styles.

- **Nitrogen Requirement:** Its low nitrogen requirement eases nutrient management in the fermentation process.

- **Competitive Killer Factor:** Lalvin QA23 is a sensitive strain, an important aspect to consider in mixed fermentation scenarios.

- **Malolactic Bacteria Compatibility:** It shows high compatibility with malolactic bacteria, enhancing its versatility for complex fermentation processes.

Lalvin K1-V1116

Lalvin K1-V1116 emerges as a particularly intriguing choice in the range of yeasts, especially renowned for its floral ester production.

Originating from the French Mediterranean region, Lalvin K1-V1116 was selected by INRA Montpellier for its distinctive qualities. This yeast is a marvel in floral ester production, especially under lower fermentation temperatures (below 16°C) with proper nutrient management. It excels in transforming compounds into an array of floral esters like isoamyl acetate and hexyl acetate, infusing meads with fresh, aromatic bouquets.

What sets Lalvin K1-V1116 apart is its resilience in challenging fermentation scenarios, including low turbidity and temperature. It's a champion in producing meads with complex aromatic profiles, making it an excellent choice for ice wines and other specialty brews.

Lalvin K1-V1116 is a go-to for meads that yearn for a pronounced floral touch. It's superb for enhancing stone fruits, such as mangoes and cherries, as well as red and tropical fruits. This yeast is a favorite for cyser production thanks to its ability to elevate the floral and fruity notes in the brew.

Specifications:

- **Fermentation Range:** With 50–95°F (10–35°C), Lalvin K1-V1116 offers versatility in brewing conditions.

- **Alcohol Tolerance:** It can handle up to 18% alcohol, suitable for various mead styles, including those with higher alcohol content.

- **Nitrogen Requirement:** The yeast's medium nitrogen need should be considered for optimal health and fermentation activity.

- **Competitive Killer Factor:** Lalvin K1-V1116 has a competitive killer factor crucial in mixed fermentation setups.

Red Star Premiere Cuvée

Red Star Premiere Cuvée yeast stands out among the myriad options, particularly for its neutrality and versatility. Red Star Premiere Cuvée, often associated with champagne, is recommended by its manufacturers for various wine brews, including reds, whites, and champagnes. Its most defining feature is its neutral fermentation profile. This means that it adds minimal yeast character to the mead, making it an ideal choice for preserving the natural flavors of your ingredients.

Red Star Premiere Cuvée excels in traditional meads, where retaining the honey's character is important. It ensures that the nuanced flavors and aromas of the honey are not overshadowed. Similarly, this yeast skillfully balances the fruit characters for fruit-infused meads, allowing them to shine in the final mead.

Given its neutral character, Red Star Premiere Cuvée is particularly suitable for meads focusing on the ingredients' natural flavors. It's a stellar choice for traditional meads, where its ability to maintain the honey's profile is invaluable. Additionally, its capacity to balance various fruit flavors makes it a good option for a wide range of fruit-infused meads.

Specifications:

- **Fermentation Range:** It offers a wide fermentation range of 45–95°F (7–35°C), providing flexibility across different brewing conditions.

- **Alcohol Tolerance:** With an impressive alcohol tolerance of up to 18%, it can cater to a variety of mead styles, including those with higher alcohol content.

- **Nitrogen Requirement:** Its low nitrogen requirement is a boon, simplifying nutrient management during fermentation.

Red Star Premier Blanc

Red Star Premier Blanc is celebrated for its versatility and neutral flavor profile, making it an excellent choice for various brews. It's widely recommended for dry whites, Cabernet, cider, fruits, meads, ports, and sodas. The yeast's ability to produce clean esters and aromas makes it a go-to choice for dry brews, where the purity of flavor is paramount.

Red Star Premier Blanc is strong in mead-making, mainly for traditional and fruit-forward meads. It excels in preserving the intrinsic flavors of honey in conventional meads and elevating the bright notes in fruit meads without overwhelming them with yeast characteristics. Additionally, it's adept at handling high-alcohol brews, though it's less suited for carbonated varieties.

Given its neutral fermentation profile, Red Star Premier Blanc is particularly suited to meads that highlight their ingredients' natural flavors. In traditional meads, this yeast can maintain the nuanced taste of honey, and in fruit meads, it can enhance the vibrancy of the fruit flavors without adding any competing yeast notes.

Specifications:

- **Fermentation Range:** Red Star Premier Blanc functions within a broad temperature spectrum of 59–86°F (15–30°C), providing flexibility for different brewing setups.

- **Alcohol Tolerance:** It has an impressive alcohol tolerance capacity of up to 16%, making it suitable for a variety of mead styles, especially those aiming for higher alcohol content.

- **Nitrogen Requirement:** The yeast's low nitrogen need simplifies nutrient management during fermentation, which benefits mead makers of all experience levels.

Voss Kveik

Voss Kveik yeast, originating from the heart of Norway's farmhouse brewing tradition, has a rich history. Sourced from Sigmund Gjernes in Voss, Norway, this yeast has been preserved and cherished through generations.

Renowned for its wide fermentation temperature range of 25–40°C (77–104°F), with an optimal range of 35–40°C (95–104°F), Voss Kveik is known for its rapid fermentation.

Typically achieving full attenuation in 2–3 days, this yeast remains consistent in flavor across its temperature spectrum, producing neutral tones with a hint of orange and citrus notes.

Voss Kveik is exceptionally suited for citrus-based meads, where its subtle fruity undertones can harmonize with and elevate the citrus flavors. Its versatility makes it an excellent choice for various mead styles, including fruit-forward and traditional meads. The ability to bring out a touch of citrus in the flavor profile makes it a go-to for innovative mead makers.

Specifications:

- **Fermentation Range:** Its 77–104°F (25–40°C) offers flexibility for different brewing environments.

- **Alcohol Tolerance:** Kveik Voss can handle alcohol levels up to 12%, catering to a range of mead styles.

- **Flocculation:** High flocculation assists in achieving clarity in the mead after fermentation.

- **Attenuation:** With an attenuation range of 76–82%, it effectively converts sugars into alcohol.

- **Nitrogen Requirement:** This yeast requires a high nitrogen level, which is essential for achieving optimal fermentation results.

Hornindal Kveik

Renowned for producing tropical flavors, Hornindal Kveik imparts notes reminiscent of stone fruits, pineapple, and dried fruit leather. This quality makes it an excellent partner for meads where a rich, fruit-forward character is desired.

One of the standout features of Hornindal Kveik is its ability to amplify ester intensity, especially when fermented at higher temperatures. This attribute is particularly beneficial for enhancing "C" hops, adding a new dimension to the flavor profile of your mead.

Given its tropical and fruity flavor profile, Hornindal Kveik is particularly suited for meads incorporating stone fruits or tropical elements. Its propensity to enrich and balance these flavors makes it a great choice for meads seeking a robust and aromatic character. While it is excellent for high-alcohol brews, it's important to note that it might not be the best fit for carbonated meads.

Specifications:

- **Fermentation Range:** With a broad fermentation temperature range of 68–95°F (20–35°C), Hornindal Kveik offers flexibility in brewing conditions.

- **Alcohol Tolerance:** It boasts an alcohol tolerance of up to 16%, accommodating a range of mead styles, especially those with higher alcohol content.

- **Flocculation:** High flocculation is a key feature that aids in achieving a clear mead after fermentation.

- **Attenuation:** The yeast demonstrates an attenuation range of 75–82%, showcasing its efficiency in converting sugars to alcohol.

- **Nitrogen Requirement:** Hornindal Kveik requires a high level of nitrogen, a crucial factor for optimal yeast health and activity during fermentation.

This has been my introduction to various yeasts that are particularly friendly for mead-making beginners. However, this is merely a starting point. The world of yeast is vast and diverse, offering *many* more options equally approachable for those just beginning to make mead. While I have selectively discussed specific yeasts to avoid giving you an *overwhelming* deluge of information, there's a whole range out there waiting to be explored.

I have prepared a table with an overview of the best yeasts for specific types of mead. This includes a few additional yeasts not previously mentioned. My intention here is not to inundate you with *every possible* yeast option available in the market—but to provide a balanced and informative guide. I want to equip you with the knowledge to make confident choices in your ingredients, without feeling swamped in the sea of possibilities!

	Cysers	Dark Berry Meads	Citrus Fruits	Traditionals	Light Fruit Melomels	Most Braggots	Bochets	White Pyments	Red Pyments	Super High ABV Brews
Lalvin QA23	X		X		X			X		
Safale US-05	X		X	X	X	X				X
Red Star Cote Des Blancs	X	X			X			X	X	
Safcider AB-1	X				X					
Lalvin D47	X	X	X	X				X	X	X
Red Star Premiere Rouge	X		X					X	X	
Kveik Lutra	X		X	X	X	X	X			X
Lalvin K1-V1116	X				X			X		X
Safale S-04				X	X	X				
Lalvin BM4x4		X							X	
Lalvin EC-1118		X		X			X		X	X
Lalvin 71B-1122		X		X			X	X	X	
Lalvin Bourgovin RC-212		X							X	
Red Star Premiere Cuvee			X	X				X	X	X
Red Star Premiere Blanc				X				X		X
Kveik Voss			X	X		X	X			X
Kveik Hornindal		X	X		X					
Mangrove Jacks M05				X			X			X
Red Star Premiere Classique				X		X		X	X	

Yeast Nutrients

Making mead requires careful attention to yeast nutrition for optimal fermentation. Honey, while rich in sugars, often falls short of providing essential nutrients, like nitrogen and phosphate. These nutrients play important roles in regulating yeast growth, fermentation pace, and the development of certain aromatic compounds. Yeast Assimilable Nitrogen (YAN), encompassing amino acids, ammonia, and select peptides, is vital in this context.

The availability of sufficient nutrients is really a make-or-break factor in mead production. Insufficient nutrients can lead to sluggish fermentation and the risk of persistent off-flavors. Interestingly, the nutrient content in honey varies: darker varieties, like buckwheat, are typically more nutrient-rich than their paler counterparts.

Diammonium Phosphate (DAP) is a nitrogen source for yeast proliferation and effective fermentation. When choosing DAP, selecting pure forms and avoiding those containing urea is crucial, as urea can contribute to undesirable flavors (like salty or metallic notes) if not fully processed by the yeast. Furthermore, urea is a precursor to urethane, a recognized carcinogen, leading to its abandonment in the wine industry. It's worth noting that many commercial yeast nutrients are a mix of DAP and urea—so label reading is key. Caution is advised against adding DAP during yeast rehydration due to the toxicity of ammonia salts in high concentrations. DAP is often used alongside Fermaid-K in mead-making to ensure a balanced nutrient supply throughout fermentation. For reference, one teaspoon of DAP weighs around 3.9 grams.

Fermaid-K, a proprietary blend from Lallemand, is another essential nutrient in mead-making. This blend provides nitrogen, crucial vitamins, nutrients, and in-activated yeast hulls. It's often labeled as "Yeast Energizer". The recommended approach for adding Fermaid-K involves two stages: initially at the end of the lag phase (about 6–12 hours post-yeast pitching), and subsequently around the point of one-third sugar depletion. This dual addition is part of a strategy known as staggered nutrient additions, a topic that will be explored in more detail in upcoming chapters. A teaspoon of Fermaid-K weighs approximately 4 grams.

Fermaid O represents a newer, advanced version of Fermaid K. Its distinguishing feature is the absence of inorganic nitrogen sources, like DAP or ammonia salts. Instead, it relies solely on organic nitrogen, which is more readily assimilated by yeast, leading to a more consistent fermentation process. Another added benefit of Fermaid O is its ability to moderate heat production during fermentation, reducing the formation of unwanted sulfur compounds that can affect flavor. Though relatively new in the mead-making scene, Fermaid O's unique properties make it a promising piece of equipment in the mead-maker's toolkit.

Within these pages in our mead-making adventure, I will shine the spotlight on using DAP and Fermaid K. It's like sticking with the classics, you know? Especially when we get into fermentation's nitty-gritty and learn about staggered nutrient additions. Sure, Fermaid O has its perks, but let's walk before we run, and master the traditional methods first. This can set a solid foundation for future experimentation.

Debunking the Myth: Raisins as Yeast Nutrients in Mead-Making

Raisins/Credit: NIPAPORN PANYACHAROEN (www.Shutterstock.c om)

Let's chat about an enduring belief and practice in historical mead-making: using raisins as a nutrient source for yeast. This idea, passed down through generations of mead recipes, suggests that a handful of raisins can provide the necessary nutrients for yeast. But we'll see that this traditional practice doesn't quite stand up to modern understanding.

As just mentioned, yeast nutrition is a critical factor in successful fermentation, and it revolves around something called yeast-assimilable nitrogen, or YAN. This includes organic sources like amino acids (Free Amino Nitrogen or FAN) and inorganic ones, like ammonia. The role of YAN is to ensure that yeast has everything it needs to ferment effectively. When we look at raisins from this perspective, their contribution to YAN is, unfortunately, relatively minimal.

Think about wine, which is made from grapes. If grapes—and, by extension, raisins—had all the nutrients needed for fermentation, winemakers wouldn't regularly add nutrients to their grape must. This common practice in winemaking indicates that grapes, and thus raisins, are *not* the nutrient powerhouses for fermentation that we once believed them to be.

You see, adding about 50 raisins to a gallon of mead is roughly 34 ppm of nitrogen—a drop in the bucket compared to the recommended 200 to 400 ppm YAN needed for robust yeast health. Moreover, not all the nitrogen in raisins is accessible to yeast, due to their high proline content—an amino acid that yeast finds hard to process.

So, while tossing a few raisins into your mead might add a hint of flavor, they're not the nutrient source you need for a healthy fermentation. For that, look towards specialized nutrients like we have seen, with diammonium phosphate (DAP), Fermaid-K, or Fermaid-O, designed to give your yeast the full banquet of nutrients it craves.

CHAPTER 3
Equipment

W hether you are starting with just the basics in equipment, or ready to look at more sophisticated gear, this guide I have written has you covered. I'll share practical insights on the necessary tools and equipment, so you're well-prepared for mead-making—regardless of your budget or level of expertise.

Basic Equipment

Plastic Fermenting Bucket

Plastic fermenting bucket

Let's discuss simple plastic fermenting buckets—key players in your mead-making activity! Their popularity among beginners and seasoned mead-makers comes down to practicality, affordability, and ease of use.

Plastic fermenting buckets are light, easy to maneuver, and have lids designed to hold airlocks, keeping your mead safe during fermentation. Moreover, their wide mouths make cleaning a breeze—a critical step for ensuring your mead is top-notch. Plus, the semi-transparent material of some buckets lets you peek at the progress inside without exposing your brew to the outside world.

Remember, your plastic bucket needs love and care. They can last for several batches if treated right, but they're prone to scratches, which can be a breeding ground for unwanted microbes. Gentle cleaning and regular inspections will help keep your bucket in prime condition. Also, consider the shape and design. The large surface area exposed to air in these buckets can sometimes increase the risk of oxidation, affecting your mead's flavor. It's a balance of convenience and vigilance.

Always choose *food-grade* plastic for your fermenting bucket to keep your mead safe and tasty. Non-food-grade plastics can be harmful, leaching chemicals into your precious brew.

In the initial stages of mead-making, a plastic fermenting bucket is usually the go-to choice for primary fermentation. This is because you'll often need to open it up to stir the must, and it provides the necessary headspace to accommodate the lively early stages of fermentation. As you progress to the secondary or finishing fermentation, where limiting oxygen exposure is critical, switching to glass carboys or jugs are ideal options.

Glass Carboys

Glass carboy/Credit: ciwoa (www.Shutters tock.com)

In mead-making practice, the design of a glass carboy truly stands out as a game-changer. Its unique shape minimizes your mead's exposure to air, significantly cutting down the risk of oxidation. This factor is vital in preserving your mead's healthy integrity and distinct flavor. Moreover, glass carboys are notably scratch-resistant, offering a hygienic solution that, with proper care, will endure.

The choice of glass for mead-making brings with it a host of advantages. Being an inert material, glass doesn't react with the mead, ensuring the flavors and aromas remain authentic and untouched. Its non-porous nature also means it won't retain any residual odors or flavors, giving every new batch a beautifully clean slate.

Another delightful aspect of using glass in the brewing process is the clarity it offers. It's fascinating (and convenient) to watch your mead develop and age within the transparent confines of a glass carboy. This visibility adds to the aesthetic pleasure—and serves a functional purpose, allowing you to keep an eye on the fermentation process, sediment layers, and the *overall clarity* of the mead, without any intrusion.

However, handling glass carboys requires mindful care. Given their fragility and weight, especially when full, they demand gentle and cautious handling. Using a carboy carrier, or ensuring a firm, two-handed grip, is highly advisable for safe transport.

Furthermore, maintaining the cleanliness and sanitation of glass carboys is paramount. Any lingering residue or contaminants can compromise the quality of your next batch of mead, making a rigorous, thorough cleaning regime essential for any successful mead-maker.

Glass Jugs

Glass jug

When crafting mead in smaller batches, glass jugs shine. These handy vessels are perfect for the mead-maker, either starting, experimenting with new flavors, or working with manageable quantities. Glass jugs are typically available in convenient sizes; such sizes are excellent for beginners testing the waters of mead-making, or seasoned brewers experimenting with new ingredients or techniques. The smaller volume of these jugs offers a more controlled brewing environment, ideal for honing your skills and fine-tuning your recipes.

Matching Your Equipment to the Size of Your Mead Batch

Mead-making begins with choosing the right size for your batch, and the right equipment. Whether you're eyeing a cozy one-gallon batch, or gearing up for a more substantial five-gallon endeavor, each batch size will have unique equipment needs. Let's walk through what you'll need for each size:

Starting Small: One-Gallon Batches

The one-gallon batch is a great starting point for those dipping their toes into mead-making or simply looking to experiment. It's compact, easy on the wallet, and doesn't require much space. Ideal for testing new recipes, a one-gallon batch typically yields about 8–10 standard bottles of mead. However, remember that if you strike gold with a recipe, you might wish to make more!

For this size, you'll need:

- Two one-gallon glass jugs, perfect for managing smaller quantities.

- A plastic fermenting bucket, preferably two gallons or larger, fitted with a lid and an airlock.

The Middle Path: Three-Gallon Batches

A three-gallon batch offers a perfect middle ground to scale up slightly. It gives you enough mead to share, while still being manageable regarding equipment size and weight. Expect to fill about 24–30 bottles from this batch. The equipment is slightly larger and a tad more costly than the one-gallon setup—but still easy to handle, with full carboys weighing about 35 pounds.

For a three-gallon batch, you should have:

- A pair of three-gallon carboys offering more volume for fermentation.

- A 6 or 6.5-gallon plastic fermenter, complete with a lid and airlock, to accommodate the increased volume.

Going Big: Five-Gallon Batches

For the committed mead maker, five-gallon batches are where things get, well, a bit serious! This size is typical for standard mead recipes and is a good fit if you want to stock up or share your creations. The equipment gets bigger here, with a full carboy weighing around 50 pounds, so ensure you have enough space.

Here's what you'll need:

- Two five-gallon glass carboys, or one five-gallon and one six-gallon carboy allow for flexibility in fermentation and aging.

- A sizable 7.9-gallon or even a 10-gallon open-top plastic fermenter, equipped with a lid, to handle the larger volume of mead.

Floating Thermometer

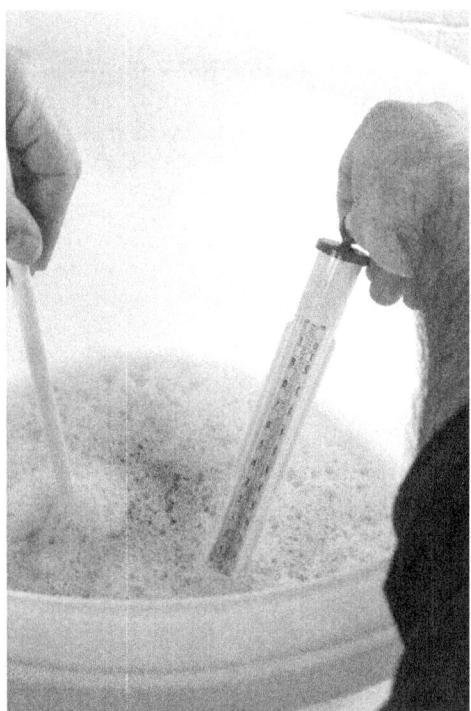

Floating thermometer/Credit: Hans Geel (w
ww.Shutterstock.com)

Temperature plays a starring role in the fermentation process. The floating thermometer is your trusty sidekick. Floating the gauge in the fermenting must will give you an accurate, real-time snapshot of the temperature, so you can check your mead is fermenting in its ideal range.

Hydrometer

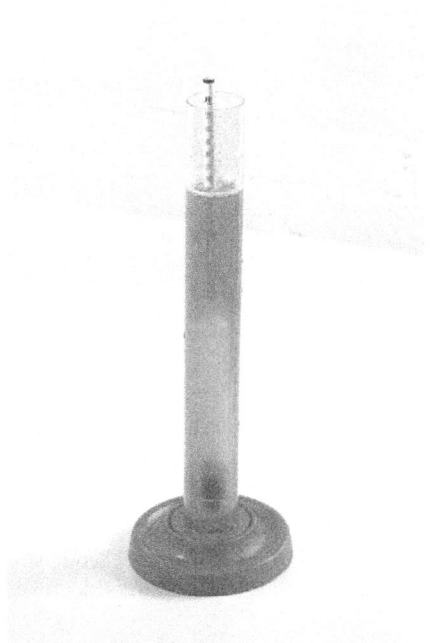

Hydrometer in a test jar/Credit: Hans Geel (www.Shutterstock.com)

As a mead-maker, familiarizing yourself with a hydrometer will be like gaining a new ally on your brewing team. This seemingly simple glass instrument, equipped with a weighted end, is vital for understanding the progress and potential of your mead.

The hydrometer's job is to measure the density of your mead mixture relative to pure water. It does this by floating in the liquid—the level at which it floats indicates the density. Pure water has a specific gravity of 1.000 at 68°F (20°C). When you dissolve sugar into the water, as in your mead must, the density increases, causing the hydrometer to float higher. This change is reflected in the specific gravity reading, giving you a direct insight into the sugar content of your solution.

The initial reading of your must's specific gravity, known as the Original Gravity (OG) or Starting Gravity (SG), is a critical milestone in your mead-making process. It's not just a number; it's a predictor of your mead's potential alcohol content and sweetness. Generally, mead SG values range from about 1.060 to 1.120, though some adventurous recipes might push these figures.

While specific gravity is the most common scale on a hydrometer, other scales like Balling, Brix, and Plato exist. These scales might present different numbers, but they all serve the same purpose: measuring the sugar concentration dissolved in your brew.

Test Jar and Turkey Baster

Turkey Baster/Credit: Matt Valentine (www.Shutterst ock.com)

Understanding the nuances of sampling your mead, while you are making it, is crucial. This is where the test jar becomes an indispensable tool. Explicitly designed for mead-makers, these jars—available in both plastic and glass—are perfect for testing small, accurate samples. They provide a convenient and efficient way to measure your mead's progress without disturbing the entire batch.

While the excitement of brewing might tempt you to check your mead often, it's essential to sample *judiciously*. Over-sampling might seem harmless, but it poses two significant risks: firstly, the more you test, the less mead you have left to enjoy! Secondly, each sampling instance carries a small risk of introducing contaminants into your precious brew.

In practice, you only need to take samples at key moments:

1. Initially, when you pitch the yeast into the must.

2. At the first break, your hydrometer readings indicate that fermentation is about one-third of the way to your target final gravity.

3. Whenever you rack your mead to a different container.

4. And, of course, right before bottling.

A turkey baster or a wine thief is ideal for extracting these samples. These tools allow you to draw out the right amount of mead for testing gently.

Airlock or Fermentation Lock

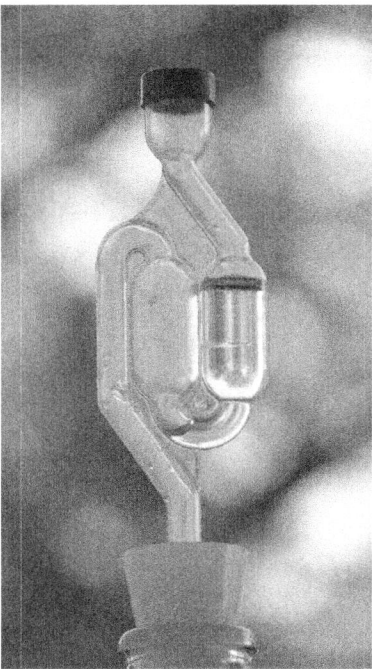

Bubbler Airlock and Rubber Stopper/Credit: Catherine H Leonard (www.Shutterstock.com)

One cannot overlook the significance of airlocks in mead-making equipment, a concept that dates back to Louis Pasteur in the 19th century. His invention of the swan-neck flask was revolutionary, paving the way for modern brewing by preventing bacterial contamination. Pasteur's work, particularly his yeast and fermentation science discoveries, is the foundation for homebrewers.

The main job of a fermentation lock, or airlock, is to form a protective barrier. It keeps out unwanted bacteria and wild yeast, ensuring your mead remains pure during fermentation. But it's not *just* about defense; airlocks also allow carbon dioxide (a byproduct of fermentation), to escape without letting in any harmful external elements.

Regardless of their design, all airlocks fulfill this vital function. They typically use a liquid barrier, be it water (or, for some, 80-proof vodka), which is preferred for its sanitizing qualities (though it does evaporate faster). As fermentation progresses, the pressure from the carbon dioxide forces it to bubble through the liquid in the airlock, effectively sealing your mead from the outside world.

Choosing the Right Airlock

3 Pieces Airlock/Credit: Hans Geel (www.Sh utterstock.com)

Among homebrewers, two styles of airlocks are popular, each with its advantages:

- **3-Piece Airlocks:** Known for their ease of maintenance, these airlocks can be easily disassembled for cleaning and sanitizing. While they're generally very reliable, care should be taken during fermentation to avoid any movement that could introduce contaminants.

- **Bubbler Airlocks:** These classic airlocks are highly effective at keeping out contaminants. A unique feature is their ability to let you visually monitor fermentation activity through the liquid levels in the chambers. However, ensuring adequate headspace in your fermenter is crucial to prevent "Krausen" from clogging the airlock (Krausen is a term used for the layer of foam that appears during the fermentation stage's height. It is a mix of yeast cells, proteins, and various compounds—all byproducts of the lively fermentation process.)

Blow-Off Tubing for Vigorous Fermentations

Carboy with a Blow-Off Tube

Certain situations call for something a bit more robust than your standard airlock in mead-making. This is where blow-off tubing can be a real lifesaver. Picture this: you attach a hose to the rubber stopper of your fermenting bucket, with the other end dipping into a jar of sanitized water. It's a simple setup yet incredibly effective, especially for those larger batches, or when fermentation takes on a life of its own. The beauty of this system lies in its simplicity—it allows carbon dioxide to escape while ensuring no unwanted guests like wild yeast or bacteria get in. Remember to submerge that tube in the sanitized water—and voilà! You've got a reliable, oversized airlock!

Crafting Makeshift Airlocks in a Pinch

Glove used as Airlock/Credit: Kartas (www.S hutterstock.com)

Sometimes, brewing throws you a curveball—you're all set to ferment, and suddenly, you realize you're out of airlocks... but fear not—you can repurpose several everyday items in these situations. While they might not match the security of commercial airlocks, they can still do a pretty decent job.

- **Aluminum Foil.** A simple yet effective solution from sanitary, food-grade aluminum foil can be made. Cover the top of your fermenting vessel with the foil, letting it drape slightly down the sides. This setup allows CO_2 to escape while acting as a barrier against bacteria and wild yeast. However, it's important to note that this method isn't entirely foolproof—oxygen can still sneak in, posing a risk of oxidation.

- **The Balloon Method.** Secure a balloon with a rubber band over the opening, ensuring it's as airtight as possible. Then, puncture a few tiny holes for CO_2 to escape. The trick here is to balance the size of the holes—they need to be small enough to prevent oxygen from getting in while allowing CO_2 to exit.

- **Other Alternatives.** In moments of need, the brewing world becomes a field of innovation. Any material that can expand to release CO_2 and minimize oxygen entry can be repurposed as an airlock. This includes using items like condoms, water-based tubing systems, or even plastic bags and kitchen wraps secured with rubber bands. While these methods can be effective, always remember the key principle of brewing: keeping everything sanitized.

Rubber Stoppers

When assembling your mead-making setup, another key component you'll need is the drilled rubber stopper. This small but essential item acts as a bridge, connecting your fermenter to the fermentation lock. The stopper fits snugly into your fermenter, with the fermentation lock then inserted into the stopper's pre-drilled hole. It's a good idea to have a couple of these stoppers on hand: one for your plastic fermenting bucket and another for your carboy. For a 5-gallon carboy, stoppers sized No. 6 to 7 are typically suitable. As for your plastic fermenter, it's wise to check with your supplier to determine the correct size.

Siphon Hose

Fermenting Bucket and Siphone Hose/ Credit:
BDoss928 (www.Shutterstock.com)

Another significant step in your mead-making activity involves carefully transferring your brew—a process known as racking. This is where the humble (yet essential) siphon hose plays its part. Crafted from clear, food-grade vinyl tubing, these hoses are indispensable for moving your mead safely from one vessel to another. While specialized brewing stores are go-to sources for these hoses, don't overlook your local hardware or home supply stores, which might offer more competitive prices. In terms of size, the most commonly used diameters for mead-making are 5/16 inch and 3/8 inch. You'll typically need a length of about six feet.

Auto Siphon

Auto Siphon

The auto siphon emerges as a game-changer in the craft of mead-making, where precision and cleanliness are paramount—especially for those who find traditional racking canes (another siphoning tool) challenging. The auto siphon tool has been widely embraced as a substantial improvement, making the transfer of mead effortless and frustration-free.

An auto siphon is a marvel of simplicity and effectiveness in design. It features a racking cane attached to tubing, encased within a larger racking tube. This setup includes a crucial filter at the tube's end, designed to catch and block any unwanted particles. A rubber grommet encircles the cane, allowing smooth, airtight movement within the tube. Despite its lightweight nature, the auto siphon's design is fundamental to its efficiency, offering a blend of basic mechanics and good effectiveness.

Its mechanism, that simplifies the siphoning process, sets the auto siphon apart. Traditional methods, which often involved creating a vacuum manually (and unsanitarily) by mouth, are replaced by an automated, cleaner approach. The auto siphon leverages atmospheric pressure and gravity, allowing for an easy start to the siphon process—without the need for potentially contaminating actions. This advancement makes the process more straightforward—and significantly enhances the sanitary conditions that are so very important in mead-making.

Sanitizer

Bleach/Credit: Steve Cukrov (www.Shut terstock.com)

It follows that sanitation is the cornerstone of successful mead-making, ensuring the safety and quality of your brew. The market offers various sanitizing solutions, each with unique characteristics and instructions. Mead-makers need to understand these options to make an informed choice that aligns with their brewing needs. Here are the most commonly used sanitizers in home brewing:

- **Bleach:** A widely accessible and cost-effective option, bleach is a popular choice among homebrewers. However, its use typically requires a very thorough rinse post-sanitization to prevent residue.

- **Metabisulfites (K-Metabisulfite and Na-Metabisulfite):** These effectively combat oxidation and remove unwanted microorganisms, making them a reliable choice for mead-makers.

- Specialized Brewing Sanitizers: Products like B Brite/C Brite, Iodophor, and Saniclean are designed specifically for brewing equipment. They offer a balance of efficacy and ease of use.

- **Star San:** Star San shines for its practicality and efficiency. Perfect for sanitizing various fermenting vessels, it can be used as a soaking solution or

in a diluted form, stored in a spray bottle for easy application. The solution remains effective for up to 28 days in a sealed container, providing a
long-lasting sanitizing option.

- **Other Popular Options:** Brewers frequently use Powdered Brewery Wash,
  Straight A, Beer/Wine Line Cleaner, and Easy Clean for their effective cleaning properties.

I personally recommend that you use Star San as a sanitizer. Star San stands out for
its user-friendly application and extended effectiveness. Utilizing it in a spray bottle
conserves the solution and ensures thorough coverage. This approach streamlines
the sanitation process, making it both efficient and economical. Its durability as a
diluted solution adds to its appeal, offering a ready-to-use sanitizing agent for weeks.

Sanitizer	Contact Time	Rinse/No Rinse	Environment Friendly?
Star San	60 seconds	No Rinse	Yes
Bleach	15 minutes	Rinse	No
Diversol	15 minutes	Rinse	Yes
Iodophor	60 seconds	No Rinse	No
B-Brite	15 seconds	No Rinse	Yes

PH Test Strips & PH Meter

Mastering the delicate balance of acidity in mead is essential. It's not just a number;
the pH level directly impacts your mead's flavor, clarity, and fermentation harmony.
To navigate these waters, mead-makers rely on pH test strips and pH meters, each
offering a window into the acidic soul of their brew.

pH Test Strips

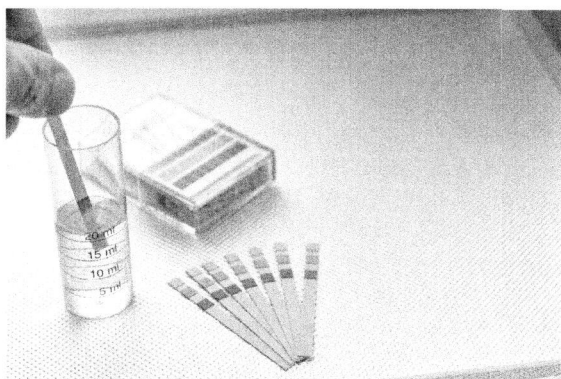

pH Test Strips/Credit: Javier Ruiz (www.Shutterstock. com)

For the mead-maker who values ease and efficiency, pH test strips are a good go-to. These humble indicators provide a color-coded glimpse into your mead's pH level. Just dip a strip into your mead, match the color change to the chart, and voilà—you have an approximation of your mead's acidity. They're a budget-friendly choice, especially when you're starting or prefer a hands-on, less technical approach to brewing.

The Precision of pH Meters

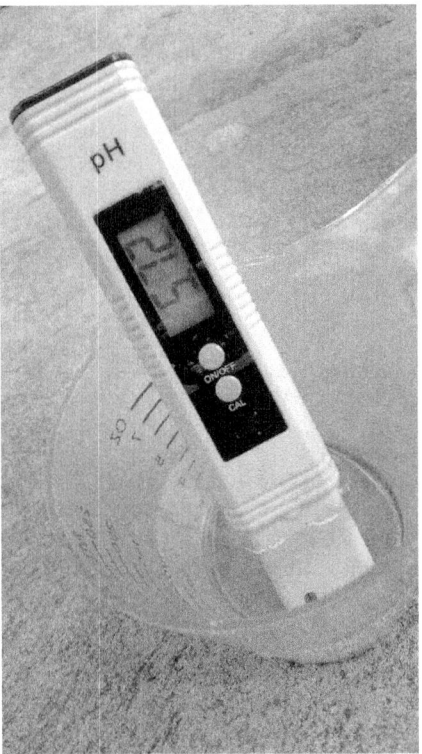

pH Meter Test/Credit: khayaw (www.Sh utterstock.com)

When precision is key, a pH meter will be your trusted companion. This gadget delivers deeper information than mere color changes, offering a precise digital readout of your mead's pH. With a calibrated probe dipped into the liquid, the pH meter provides an accurate and immediate understanding of your mead's condition. For the dedicated mead-maker who seeks consistency and control, the pH meter is an investment in your craft.

Selecting between pH test strips and a pH meter hinges on your preferred brewing practice. If you enjoy simplicity and quick checks, pH test strips are your ally. For those who revel in the details and aspire to really fine-tune their process, a pH meter becomes an indispensable tool in your mead-making repertoire.

Everyday Kitchen Tools

In addition to this specialized equipment for mead making, you'll also need to collect a few standard accessories that are typically found in most kitchens:

- A large bucket and lid in which to prepare your sanitizing solution (water and your sanitizer of choice)

- A large container with a lid (it should be able to contain at least as much liquid as the amount of water you are planning to use for your mead) for storing and treating your water before pouring it into the fermenter.

- A kitchen spatula to efficiently gather any remaining honey from its container

- A large measuring cup

- A set of measuring spoons

- A resealable plastic bag for mixing the yeast nutrients

- A small measuring cup, preferably in glass or plastic, for yeast rehydration

- A small spoon for stirring the yeast during rehydration

- A plastic film wrap to cover the yeast as it undergoes rehydration

- Standard kitchen glasses

- Any digital time-tracking device

- Scissors

Bottling Equipment

Having delved into the fundamental tools needed for crafting mead, our next step is to turn our attention to the bottling phase. This is where your hard work comes to fruition, and the right equipment is necessary.

Bottles

Counting Bottles

Bottles/Credit: rozbeh (www.Shutterstock.com)

As your mead's fermentation time ends, it's time to think about bottling—the final chapter in creating your liquid gold. You'll need about ten 12-ounce bottles for a single gallon of mead, neatly holding every drop. Upscale to a three-gallon batch, and you're looking at 30 bottles, a satisfying collection for any mead enthusiast. And for those grand five-gallon ventures, 50 12-ounce bottles will be the vessels of your labor. If 16-ounce bottles are more your style, eight will suffice for a gallon, 24 for three gallons, and 40 for a five-gallon batch, each one a larger quantity for your mead's expression. For those with a flair for the ceremonious—Champagne-style bottles add a touch of elegance, with about five needed for a gallon, 15 for three gallons, and 25 to cradle a five-gallon batch in their graceful curves.

Finding the Perfect Bottles

You can buy brand-new bottles from your local homebrew shop, but if you're feeling resourceful, why not collect and recycle? If you go down the collection route, remember quality matters. Seek out bottles meant for multiple uses—those robust longnecks without twist-off threads are ideal, especially for still mead. You need bottles that can handle pressure for the sparkling variety, like those used for American sparkling wine. These can take a crown cap and will stand up to the effervescence of your sparkling mead without a fuss.

Once you've gathered your bottles, cleanliness is your next mission. A simple, thorough rinse right after use will set you up for success when bottling day arrives. Ignore this step, and you might end up with a colony of unwelcome growths to clean out, turning bottling day into a scrubbing marathon!

When it's time to decide on bottles, consider the type of mead you've crafted. A still mead is accommodating and doesn't demand much from its container. On the other hand, a sparkling mead, with its zest and vigor, needs a stronger bottle that can stand up to the pressure—think sturdy beer or reliable Champagne bottles.

Light exposure is mead's subtle foe, affecting flavor over time. If your mead will sit in the light, shield it in green or brown bottles. Feel free to use clear bottles if it's destined for a dark cellar or cabinet. When planning long-term storage, the seal is critical—corked, capped, or flip-top bottles are your best allies. If you're considering screw-top bottles, avoid those designed for single use unless you have the appropriate capping tools.

How you enjoy your mead should also influence your bottle selection. If your practice is to savor a whole bottle in good company, wine bottles may be your choice. Beer bottles offer the perfect portion for those who enjoy mead in moderation. And remember—nothing stops you from keeping various sizes on hand, to cater to every occasion.

Caps & Corks

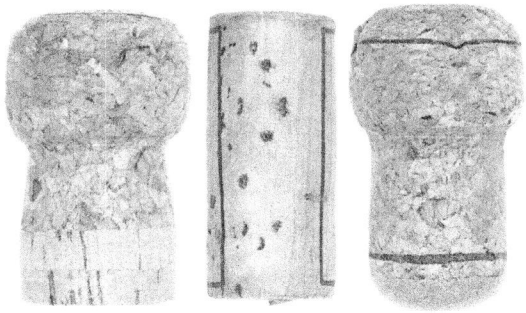

Bottle Corks/Credit: eNjoy iStyle (www.Shutterstock.c om)

Corks, the "guardians" of your mead's essence, come in various forms: natural, composite, and plastic. Each type has unique benefits, but all serve the necessary purpose of sealing in the flavors and aromas of your mead. The key is to select a cork that matches the bottle's size, with #9 being a standard choice for many wine bottles. But the selection is just the beginning. Before insertion, ensure that each cork is thoroughly sanitized. This step is essential to preserve the purity and quality of the mead you've rather painstakingly crafted.

The selection process for mead-makers who prefer bottles with caps is pretty straightforward. Caps are readily available at any homebrew store and are generally standard in size and style. They offer a simple and effective way to seal your bottles, protecting your mead from external elements. However, it's important to note that *not all* bottles are suited for caps. Specifically, twist-off bottles are a no-go—they don't provide the necessary seal for long-term storage. Using such bottles risks the integrity of your mead, possibly leading to oxidation and a very compromised flavor profile.

Bottle Filler

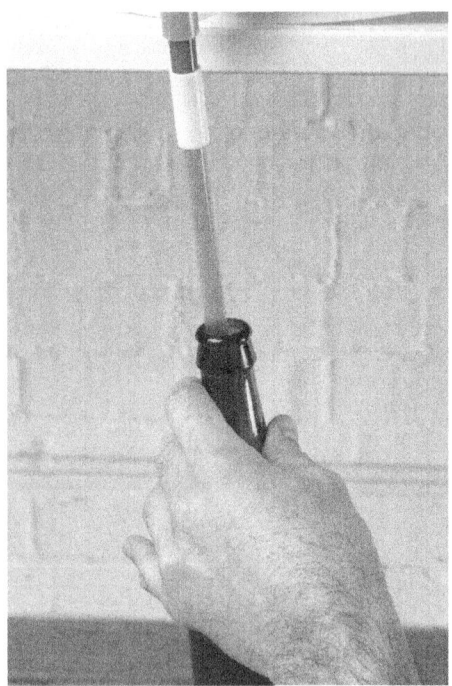

Bottle Filler/Credit: Hans Geel (www.Shutterstock.com)

This ingenious tool, a simple yet transformative tube, attaches effortlessly to your siphon hose, turning the (often tedious) task of filling bottles into a seamless and quite enjoyable process.

At the heart of a bottle filler's utility is its shut-off valve, ingeniously designed to control the flow of mead. There are two types of valves commonly used: spring-actuated and gravity-actuated. Each type has strengths, but the spring-actuated variant is particularly adept at minimizing drips and spills. As you insert the filler into a bottle, the valve opens, allowing mead to flow. Once the liquid reaches the desired level, lift the filler, and the valve ceases the flow, leaving just the right amount of "headspace" for corking or capping. This precise control over the flow ensures uniformity across bottles and keeps your workspace clean—and mead loss to a minimum.

Choosing between a spring- or gravity-actuated bottle filler often comes down to personal preference and the specific requirements of your bottling setup. The key is to find a filler that complements your bottling technique, ensuring a smooth, efficient, and drip-free process. Whether you're a seasoned mead-maker—or just starting out—integrating a bottle filler into your process is a step towards mastery in mead-making.

Bottle Corkers and Cappers

Choosing a bottle corker becomes necessary when your mead reaches its final stage. Corkers are categorized into three types: wing, tabletop, and floor models, each suited to different needs (and budgets).

- **Wing Corkers:** These are the most cost-effective, typically available for around $22–25. Their compact design makes them easy to store, an ideal choice for those with limited space. However, they require a careful touch to avoid damaging the bottleneck—a small trade-off for their affordability.

- **Tabletop Corkers:** A step up in price and stability, tabletop corkers, priced at about $65–70, offer a more user-friendly experience. They are slightly bulkier, but their design allows easier corking with less risk to the bottles.

- **Floor Corkers:** The premium choice in corkers and floor models, priced at just over $100, the floor corker provides the most stable and effortless corking experience. Their cost is higher, but for avid mead-makers, the investment can be well worth the ease and safety they offer.

Wing Corker/Credit: Melih Evren (www.Shut terstock.com)

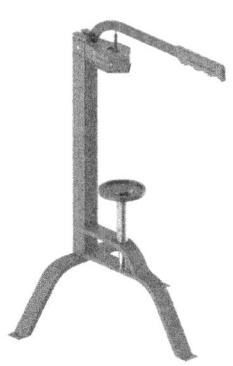

Floor Corker

Regarding cappers, two main types are prevalent: two-handled and bench cappers.

- **Two-Handled Cappers:** These cappers are a practical choice with an affordable price range of $13–17. They require a bit more effort but are straight-forward in their use.

- **Bench Cappers:** For those looking for better efficiency, bench cappers, which cost between $40–60, are a great option. They speed up the capping process, allowing you to hold the bottle with one hand and cap with the other.

Two-Handled Capper/Credit: Steve Heap (www.Shutt erstock.com)

Bench Capper/Credit: Roger Siljan-der (www.Shutterstock.com)

If purchasing a corker or capper seems like a significant investment, especially for those who bottle infrequently, renting may be a sensible alternative. Many homebrew shops offer these tools for rent, providing a cost-effective solution for your occasional bottling needs.

My recommendation leans towards a floor corker and a bench capper for consistency and quality in your mead. Their higher price point is justified by the level of precision they offer. Improper sealing can lead to oxidation, and the awful compromising of your mead's quality. (If buying these tools outright is not feasible, consider seeking rental options—or borrowing from fellow mead enthusiasts!)

Bottle and Carboy Brushes

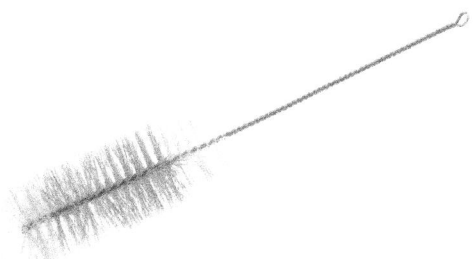

Bottle Brush/Credit: Alexthq (www.istockphoto.com)

Bottle and carboy brushes are tools designed specifically for the varied sizes of bottles and carboys used in mead-making, ensuring that each kind of vessel is meticulously cleaned.

The variety of brushes available is tailored to meet the diverse needs of mead-makers. They come in multiple sizes, perfectly matching the dimensions of different bottles and carboys. This ensures that whether you're scrubbing a petite bottle or tackling the larger surface area of a carboy, there's a brush that fits like a glove, making your cleaning process thorough and effective.

In addition to size variety, these brushes also come in different shapes. The straight brushes are superb for general cleaning, efficiently scrubbing most interior surfaces. However, the unique angled brushes are the unsung heroes in this cleaning process. They are specially designed to reach the elusive spots, particularly around the tops of bottles and carboys where residues often hide. This flexibility in design is a big plus, as it allows for a comprehensive cleaning regimen, leaving no spot untouched.

Advanced Equipment

Funnel

Funnel/Credit: Ispace (www.Shutterstock.com)

The importance of a reliable funnel cannot be overstated in the art of mead-making, especially when working with carboys or jugs for primary fermentation. The challenge of pouring liquid into the narrow neck of a carboy is not only in avoiding spills, but also about ensuring safety—mainly when dealing with hot must.

When choosing a funnel for your mead-making activity, its ability to withstand high temperatures is a crucial consideration. The funnel must be resistant to heat and capable of handling the freshly boiled must without any risk of warping or damage. This is vital, not just for the funnel's longevity, but also for *your safety* and the integrity of your mead. A warped or damaged funnel could lead to spills or accidents, jeopardizing both safety and integrity.

Some homebrew stores offer specialized funnels designed for brewing that are equipped with an added feature: built-in filters. These filters are a blessing to the mead-maker, allowing you to effortlessly remove any unwanted solid ingredients from the must as it's poured into the fermentation vessel. This feature enhances your mead's clarity and quality, streamlining the brewing process and ensuring a smoother, purer final product.

Yeast Starter Bottle

Yeast Starter Bottle/Credit: jannoono28 (www.Shutte rstock.com)

When preparing your yeast starter, the choice of container is essential. While any sanitizable container can suffice for rehydration, a vessel accommodating an airlock is ideal for a yeast starter. This setup minimizes the risk of contamination while allowing the yeast to flourish. Glass or plastic containers are typically preferred due to their non-reactive nature. If you opt for metal, avoid aluminum, as it can react with the starter solution, potentially impacting the quality of your yeast.

Measuring Spoons, or a Small Scale

Measuring Spoons/Credit: Hurst Photo (www.Shutter stock.com)

Measuring spoons allow you to replicate your recipes precisely, ensuring consistency in every batch of mead you create. For mead-makers passionate about precision, using a small scale can further elevate the accuracy of your measurements. While measuring spoons are great for general use, when it comes to ingredients that require exactitude, especially in small quantities, a scale measuring fractions of a gram becomes an invaluable tool. This is particularly true for ingredients like nutrients, where even slight variations can impact the fermentation process, and the final flavor of your mead.

Large Scale

Large Food Scale/Credit: v74 (www.Shutterstock.com)

Especially when it comes to measuring foundational ingredients like honey and water, a scale that can handle up to 25 pounds (or 12 kilograms) and displays measurements in fractions of a pound or kilogram, can significantly enhance the accuracy of your proportions.

A large scale, often found in restaurant kitchens, is ideally suited. It allows you to measure large quantities of honey with almost the same exactitude as more minor ingredients.

Agitator Rod/Wine Degasser or Wine Saver

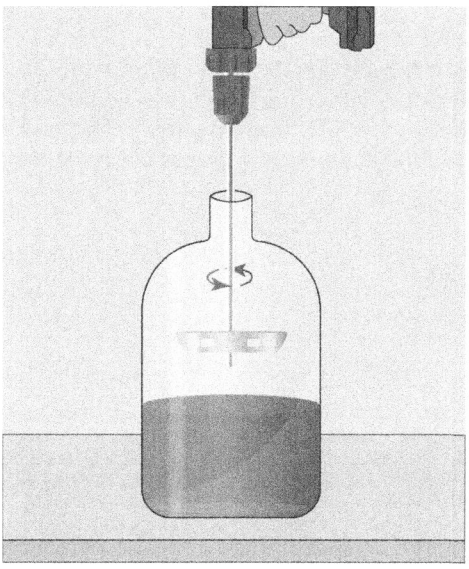

Wine Degasser

Understanding the specific equipment to support processes like oxygenation and degassing cannot be overlooked in our mead learning phase! While the next chapter will detail these processes, we will introduce you here to the essential tools that significantly impact oxygenating and degassing your mead.

When I began my mead-making adventure, I thought it was a straightforward process of mixing ingredients and waiting! However, I quickly learned that achieving the perfect mead involves a much more attentive approach, especially during the degassing phase. This phase is critical because, during primary fermentation, yeast converts oxygen into CO_2. While much of this CO_2 naturally escapes during fermentation, some remains, necessitating manual removal.

It's important to distinguish between oxygenating and degassing. Oxygenation is vital at the onset of fermentation, supplying yeast with the oxygen needed for the successful fermentation process. However, introducing additional oxygen post-fermentation can be detrimental. The remaining yeast will continue to produce CO_2, and excess oxygen might lead to undesirable byproducts in your mead.

Despite being distinct processes, oxygenation and degassing require similar tools. These tools are designed to introduce oxygen at the beginning of fermentation and remove excess CO_2 afterward.

- **Manual Stirring Implements:** A simple wooden spoon can be enough to stir and release CO_2 for smaller batches of mead. However, in larger batches, manual stirring can become physically demanding. This is where tools like agitator rods or wine whips, which can be attached to a drill, prove invaluable. They are ideal for degassing large volumes of mead efficiently and with less effort.

- **The Wine Saver Advantage:** Another effective tool for degassing is the Wine Saver. Initially designed for preserving opened wine bottles, it creates a vacuum to pull out air and CO_2 from the mead. This tool is particularly effective because it can effortlessly extract excess CO_2, ensuring a thorough degassing process.

Silicone Water Bungs

Silicone Water Bungs

These bungs, crafted from high-quality, food-safe silicon, present an intriguing alternative to traditional airlock systems. Their use in mead-making warrants a detailed exploration of their merits and potential drawbacks compared to standard airlocks. Below are some of their advantages:

- **Simplicity and Maintenance:** The user-friendly nature of silicon waterless bungs stands out. They are easy to handle, inserting and removing with minimal effort. Cleaning and maintaining these bungs is straightforward, requiring only a routine wash and sterilization, which adds to their appeal.

- **Durability and Sustainability:** Constructed from durable silicon, these bungs are built to last and offer extensive reusability. This durability translates into long-term cost-effectiveness.

- **Enhanced Protection Against Contaminants:** One of the standout benefits of silicon waterless bungs is their superior seal, significantly lowering the risk of contamination. This aspect is particularly advantageous over traditional airlocks, where the risk of contamination can increase if the water in the airlock evaporates or becomes polluted.

Considering the Drawbacks:

- **Visual Indicators of Fermentation:** Unlike traditional airlocks that provide visual feedback through bubbling during active fermentation, silicon bungs lack such indicators. This absence can pose a challenge, particularly for novice mead-makers, as it requires alternative methods to monitor the fermentation's progress.

- **Managing Pressure Build-Up:** Silicon bungs might not release gas as effectively as airlocks in cases of vigorous fermentation. This potential for pressure build-up necessitates vigilant monitoring to avoid any unexpected fermentation issues.

The decision to use silicon waterless bungs over traditional airlocks boils down to balancing their practical benefits and the need for attentive fermentation management. While silicon bungs offer ease of use, durability, and enhanced sanitation, they also demand a more hands-on approach to monitoring fermentation progress and pressure.

Refractometer

Refractometer/Credit: Svarun (www.Shutterstock.co m)

At the heart of mead-making lies the challenge of measuring specific gravity accurately—a task traditionally handled by the hydrometer. Its affordability and functionality make it a staple in the mead-maker's toolkit. However, this tool's precision is highly dependent on temperature. A hydrometer is calibrated to a specific temperature;

any variation can throw off your measurements. A too-warm mead might appear less dense than it is, leading to potentially incorrect conclusions about its sugar levels. It's crucial, then, to adhere to the hydrometer's guidelines, adjusting your readings for temperature to ensure the fidelity of your results.

Another practical concern with hydrometers is their appetite for ample samples. To gauge the specific gravity, you'll find yourself pulling a fair amount of mead into a cylinder—mead that you will often pour away to prevent contamination. This practice (especially if you're the type to check gravity diligently) can add up to a considerable amount of mead *wasted* over time.

When you open your fermenter, you roll out the red carpet for potential contaminants. This risk escalates if you're trying to bypass the sample wastage by leaving a hydrometer in the fermenter—a gamble that could *endanger the entire batch.*

Using a hydrometer isn't a one-man show; it's an ensemble that demands a collection of supporting tools. A thief is necessary for drawing samples; a graduated cylinder for reading them; and a thermometer for adjusting to temperature variances. While owning a hydrometer is a step towards precision, it's just part of a more extensive set of tools for accurate measuring.

Refractometers represent a significant advancement in measuring sugar content in the meticulous art of mead-making. Unbound by the constraints of liquid density, refractometers calculate sugar levels by assessing how light bends through the mead, offering a new dimension to brewing precision.

Here are some of the advantages of using a refractometer over a hydrometer:

- **Conservation of Mead:** The refractometer's need for only a droplet-sized sample is a testament to its efficiency. This means more of your carefully crafted mead remains in the fermenter—destined for the final bottling rather than waste!

- **Mitigating Temperature Variables:** Rapid cooling of small samples aligns perfectly with a refractometer's design, rendering the instrument less vulnerable to temperature shifts. Many models boast built-in temperature compensation, neutralizing minor discrepancies that could otherwise skew readings.

- **Streamlined Toolset:** The refractometer's process simplifies your toolkit. Gone is the need for a bulky wine thief or turkey baster, and a graduated cylinder. A sterilized dropper or pipette suffices for sampling, exemplifying the refractometer's streamlined approach.

- **Reduced Risk of Contamination:** The refractometer minimizes the need to expose the fermenter. A swift dip of a pipette through the airlock's gap significantly lessens the risk of introducing contaminants, as opposed to the wider breach required by traditional methods.

With a refractometer, the sampling frequency can safely increase, allowing you to track your mead's fermentation journey more closely. This frequent data collection can alert you to early deviations—enabling timely interventions.

The refractometer also fits seamlessly into various stages of the brewing process, from checking pre-boil gravities to monitoring mash efficiencies, with little concern for fluctuating temperatures or large sample requirements.

While a hydrometer may suffice for beginners, or those mindful of expenses, the refractometer is a good investment in your craft's future. For those who have moved beyond the basics, its value lies in its convenience, consistent precision, and insight into your mead.

Digital Hydrometers: The New Frontier in Mead-Making

As we chart the course of mead-making into more sophisticated territory, the digital hydrometer also stands as a beacon of modernization. With its precise electronic sensors, the digital hydrometer offers an immediate digital readout, eliminating the guesswork associated with analog scales. This leap forward enhances accuracy and simplifies the brewer's task. No longer must one squint at a submerged scale; a precise, unequivocal number now presents itself, making precision in measurement accessible to all!

The Benefits:

- **Continuous Tracking:** The standout feature of many digital hydrometers is their ability to offer real-time monitoring. This constant insight into the fermentation process negates the need to disturb the mead, preserving its purity while keeping you informed.

- **Adapting to the Environment:** With built-in mechanisms for automatic temperature adjustment, digital hydrometers deliver accurate readings without the fuss over environmental fluctuations—a true ally in the unpredictable realm of home-brewing.

- **A Chronicle of Fermentation:** The capacity to log data over the entire fermentation period allows for a retrospective analysis of each batch. This historical data can be a treasure trove for refining techniques and achieving consistent results.

Other Considerations:

- **Investment for the Future:** While the upfront cost of a digital hydrometer may be higher than that of a traditional one, the investment speaks volumes about your commitment to quality and innovation in mead-making.

- **Keeping It Precise:** A digital hydrometer demands regular calibration to ensure its precision endures. Proper maintenance is also essential to safeguard the electronic components that are the lifeline of its functionality.

CHAPTER 4
Basic Mead–Making Process

W ithout further ado, here is your start in the fascinating world of mead-crafting, where we impart the foundational knowledge and practical steps you will take to turn honey into the liquid gold revered by cultures across the globe. This introduction is designed to demystify the process, breaking it down into manageable, bite-sized chunks and ensuring that even beginners feel empowered to start! Whether you want to make a batch or two to share with friends, or further indulge your curiosity about this ancient beverage, you're in the right place here!

The Art of Record-Keeping in Mead-Making

Think of it this way: every batch of mead is a story. And like any good storyteller, you need notes to keep the narrative consistent and engaging. When you craft a mead that's just *perfect,* wouldn't you want to relive that success? That's where your records come in—they're your secret recipe book. But it's not just about repeating triumphs; it's also about learning from the batches that *didn't quite* hit the mark. Your records are your detective kit, helping you pinpoint exactly where to improve.

Keeping records doesn't mean just scribbling down ingredients. Oh no—it's much more than that. We're talking about documenting every single step of your mead's journey. Remember when you stirred the batch at midnight, or added those nutrients? *Write it down.* These small details can make a *big* difference. And don't forget to note the temperatures and specific gravities; these are like the vital signs for the body of your mead!

And here's something else—your senses. Your notes should capture how the mead smells and tastes during its development. These personal observations are invaluable; they make your mead uniquely *yours.* Also, keep track of how you store your mead. Did you shift it from a warm corner in your room to the cool basement? Every little change in aging conditions can influence the flavor profile of your mead.

To make all this easier for you, I've got a little something up my sleeve... it's a Mead-Making Tracker designed just for mead-makers like you. It'll guide you in noting down every little but important detail. You can download it using the QR code provided. This little tool will be a game-changer for you on your mead-making journey.

Step 1: Sanitation

Let's look at a pretty crucial element of mead-making—sanitation. Picture this: after weeks of anticipation, you find yourself pouring *gallons* of your mead down the drain... heartbreaking... This scenario underscores the mantra every mead-maker should live by: "Sanitize Everything!" It's the golden rule that can *make or break* your brewing journey.

Your Brewing Space

First things first, *where* you brew matters. Often, it's the kitchen that becomes your mead-making haven. But remember, it's not just about the space; it's about how you prepare it. Close those windows, especially on breezy days. Are fans blowing towards your brewing area? Turn them off. You're very important in keeping unwelcome airborne guests away from your precious mead. And vinegar, while great in salads, is not a friend here—store it away to avoid accidental vinegar-making adventures. A thorough clean-up of your space, allowing dust to settle, sets the stage for a successful brewing session. Ever thought about the sneaky gaps under doors? A simple towel can be a barrier against dust and dirt. Pets are adorable—but let's keep those furry friends out of your brewing zone. (They might not appreciate your efforts, but your mead certainly will!)

Cleanliness Starts with You

What about you: the brewer? Your cleanliness is as important as that of your equipment. Think of yourself as part of the brewing environment. Wear clean, snug clothes, and keep your hair out of the way. Clipped, scrubbed fingernails might seem trivial, but they *matter*. Some mead-makers go the extra mile with a pre-brew shower—it's all about minimizing any risk of contamination. And don't forget, a good handwash, or a dab of hand sanitizer is a good final touch before you start.

Surface Matters: The Last Line of Defense

Your brewing surfaces are your last line of defense against unwanted microorganisms. You must keep them as *clean* as the rest of your setup. It's a lesson many have learned the hard way—even a freshly wiped countertop doesn't guarantee safety if you let your guard down and rest unsanitized equipment on it.

Timing Is Key: Sanitizing Equipment

When cleaning your equipment, timing isn't just a detail—it's essential. Start about 15 minutes before you begin brewing. Why 15 minutes? It's the sweet spot that ensures everything is sanitized just right—not too early to let contaminants resettle, and not too late that you're rushing against the clock.

Preliminary Equipment Check

Sanitation isn't just a step in mead-making; it's really an art. It begins with a simple yet crucial task: ensuring your equipment is quite spotless. Imagine this: any tiny bit of leftover gunk can be a playground for bacteria and molds, threatening to spoil your must. The solution? A thorough scrubbing session. Grab that bottle brush, armed with a generous squirt of detergent and hot water, and give your carboys, buckets, bungs, airlocks, tubing, and racking canes a good, rigorous clean! The goal is not to leave a *single* trace of residue or any lingering stale odors.

Now, let's talk plastics. They're a bit tricky, aren't they? You see, plastic containers can get scratched easily, and these scratches become cozy hideouts for bacteria. So, if you're cleaning a plastic fermenter, go gently. Use soft brushes or sponges; remember, it's not about how hard you scrub, but how effectively you do it. And for those stubborn spots? Let them soak. Sometimes, letting your equipment bathe in hot water overnight does the trick.

Sterilization of your Equipment

Sterilizing your brewing gear is like setting up a secure, clean foundation for your mead. True, achieving 100% sterilization at home is a tall order, but we aim to get as close as possible. So, how do we do it?

Heat: The Natural Sterilizer

Let's start with heat—it's nature's way of sterilizing. You have two choices: boiling your equipment for at least a minute, or using the pasteurization method, where you heat water to a certain temperature and hold it there to say goodbye to any unwelcome microorganisms. But here's a pro tip: be mindful of what you're boiling. Plastics, for instance, don't take to high heat kindly. Items like bungs and airlocks can usually plunge into boiling water, but siphon tubing? Not so much.

For the best results in your mead-making adventure, I'd personally advise leaning towards chemical sterilization for your equipment, instead of using heat. This method offers a more practical and efficient approach, and ensures higher precision in eliminating unwanted microorganisms.

Chemicals: The Science of Clean

You may have already heard about using household bleach for sanitizing brewing gear. While it's a common suggestion—it's not one I'd encourage. Here's why: post-bleach rinsing is a must—but it's like opening your doors to wild yeast and bacteria, the exact foes we're trying to keep out! Plus, if you don't rinse thoroughly, you risk creating chlorophenols. These chemicals are notorious for their unpleasant smell and taste and can seriously affect the quality of your mead. So, bleach might seem handy, but it's a double-edged sword.

Let's turn our attention to the modern marvels of sanitization. Among the various options, I'm a big fan of Five Star's Star San. This acid-based sanitizer is a game-changer for several reasons. Firstly, it's super quick-acting, and you don't need to rinse it. Yes, you read that right—*no rinsing*! This means you're not risking any contamination post-sanitization.

And what about that foam it leaves behind? Well, that's no cause for concern. It's better to leave it be. After draining, the tiny amount of Star San left won't affect your mead's taste or aroma. That's a relief!

So, how do you use Star San effectively? It's pretty straightforward. The product usually comes in a handy container with a built-in measurer. I like to mix up about 3 gallons (11.35 liters) of sanitizer in a bucket and keep it ready. Cover the bucket to keep dust out, and you're all set. Always follow the instructions on the package for the right concentration. For Star San, it's generally 1 ounce (30 milliliters) of concentrate for five gallons (19 liters) of solution. If you're mixing up 3 gallons, you will use about 0.6 ounces (18 milliliters).

Your water type matters, too. If you have a water softener at home, use the softened water for mixing. The solution stays effective as long as it's clear and has a pH of 3.0 or lower.

And when it comes to applying Star San, there's no need for soaking or filling. Just ensure the surface is completely wet for about 1–2 minutes. I find keeping a small spray bottle of Star San super handy. It's perfect for a quick spritz on carboy stoppers or airlocks, and it's much easier than sloshing solution around in larger fermenters.

Preparing the Sanitizer Solution

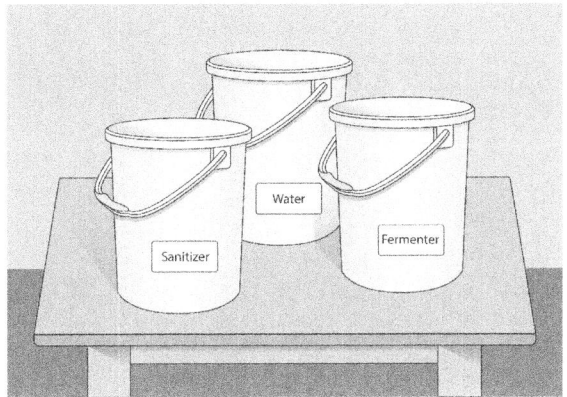

3 Buckets Required for Mead Making

Let's start with preparing your sanitizer solution.

- Mix 0.6 ounces (18 milliliters) of Star San concentrate with 3 gallons (11.3 liters) of tap water in a large, covered bucket (container 1). If your home has a water softener, it's best to use the softened water. Soft water keeps the solution effective for a more extended period than hard water. Remember, this sanitizer solution is quite efficient—it only needs a couple of minutes of contact time to do its job. There's no need to fill your containers; ensure the inside is wet with the solution. One of the great things about this sanitizer is that you can store it and reuse it. Just keep it covered in the bucket to prevent any dust or lint from getting in.

- Begin by pouring a few cups of the sanitizer solution into a large, lidded container (container 2). This container is key, as it's where you'll prepare the water for your mead. Give the container a slosh, ensuring the sanitizer reaches every nook and cranny, (lid included). Choosing a container big enough to hold the volume of water you'll need for your brew is crucial. After you've coated the inside thoroughly, return the sanitizer to your storage bucket.

- After sanitizing, you'll fill the large container (container 2) with the tap water you need to prepare your mead, according to the recipe you follow, and your desired size mead batch. Then, you'll use a clean, dry spoon to crush a Campden tablet inside the measuring cup, then add it to the water container. Give it a good stir to make sure it's fully dissolved.

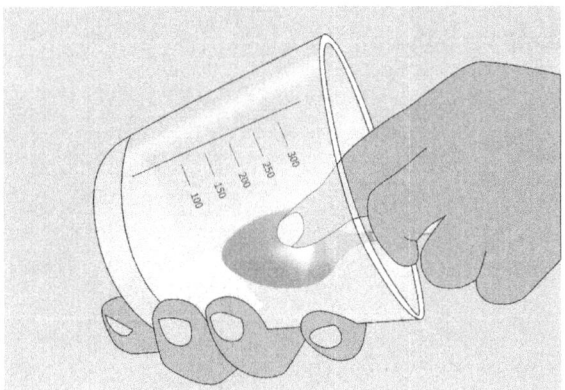

Crushing a Campden Tablet

- Temperature control is crucial. You want the water to be between 65–75°F (18–24°C). If it's not in this range, cover the container and let it sit until it reaches the desired temperature. Once done, return the thermometer to the sanitizer.

- Next up is your large plastic fermenter (container 3). Pour a few cups of sanitizer into it, swishing it around to cover every inch of the inside. Pour the excess back into the bucket. Remember to sanitize the inside of the fermenter's lid, too.

- Finally, don't forget your other tools, like the large spoon, kitchen spatula, measuring cup, and hydrometer. Place them in the sanitizer individually, making sure to splash sanitizer over all surfaces—including the handles.

Step 2: Must Preparation: Mixing Honey and Water

Among the myriad of prep techniques available, three shine brightly as the beacons for aspiring mead-makers: the Boil, Pasteurization, and No Heat methods. Each path offers its own set of scenic views and challenges, making the choice a personal adventure. Let's start with the Boil method, a method steeped in tradition and favored for its historical roots.

Boil Method

Boiling Must/Credit: BDoss928 (www.Shutterstock.co m)

The Boil method is like the old, wise sage of mead-making techniques. It's a testament to the time-honored practices of our ancestors, who crafted mead with what they had on hand, under the guidance of centuries-old wisdom.

Why Boil?

- **Clean and Clear:** By boiling, you're giving your must a clean slate, free from unwanted guests. This process also skims off the less desirable elements, like wax and pollen, leaving behind a clearer, more refined liquid.

- **Dissolution Mastery:** Boiling ensures that every drop of honey melds seamlessly into the water, creating a perfectly uniform mixture.

However, every rose has its thorns, and the Boil method is no exception:

- **Cooling Woes:** The journey from the boiling pot to the fermenter is fraught with peril—cooling must be managed carefully to avoid inviting contaminants.

- **Beware the Burn:** Honey can be temperamental under heat. A moment's distraction, and you might find yourself with a batch of caramel, rather than mead!

- **The Essence Escape:** Boiling can be too zealous, driving off the delicate aromas that give honey its soul.

- **Nutrient Exodus:** Along with the scum, you might find yourself skimming off some of the very elements that contribute to the richness and complexity of your mead.

Fear not, for these hurdles can be gracefully cleared with some savvy, as listed here! Cooling can be expedited by boiling just a fraction of your water with the honey, and introducing it to its cooler brethren already in the fermenter. This not only quickens the cooling process but also guards against contamination. Stirring is your friend during the boil, keeping the honey in motion, and away from the dangers of direct heat.

Step-by-Step Process

- Heat your water to a rolling boil.

- As you introduce the honey, stir like you're conducting an orchestra, ensuring it dances in the heat without touching the bottom.

- Boil for about 15 minutes, skimming off unwanted guests floating to the surface.

- Cooling is your next challenge. Remember, patience and cold water are your allies here.

Never, under any circumstances—pour your boiling must into a glass fermenter. The shock of the heat can lead to cracks, breaks, and, worst of all—burns. Always allow your must to cool sufficiently, or use the cold water method to safeguard against accidents.

Pasteurization Method

This technique provides a refined touch to the preparation of the must, balancing the need for cleanliness with the desire to preserve honey's intricate flavors and aromas. Unlike the more rigorous Boil method, Pasteurization gently warms the must to a temperature unfriendly to unwanted micro-organisms—but kind to the delicate nuances of the honey.

Pasteurization hinges on the principle that most yeast and bacteria find survival challenging at high temperatures, especially over extended periods. However, this method doesn't push the temperature to boiling point. Instead, it finds a sweet spot where harmful bacteria are neutralized, and the precious characteristics of the honey remain largely intact.

Why Pasteurize?

- **Efficiency in Dissolution:** The process ensures honey is perfectly blended into the water, making for a smooth and uniform must.

- **Flavor Preservation:** Avoiding a full boil retains many volatile components contributing to honey's flavor and aroma.

- **Quicker Cooling:** The must cools down more rapidly than boiling, streamlining the process.

- **Clarity:** The denaturing effect on honey's proteins during Pasteurization aids in achieving a clearer mead, as these proteins are among the culprits that can cause cloudiness.

- **Impurity Removal:** Similar to boiling, this method also facilitates the removal of impurities like bee parts and pollen, which rise as scum and can be skimmed off.

Despite its advantages, the Pasteurization method does not come without its drawbacks:

- **Loss of Complexity:** While it's more gentle than boiling, some of the honey's varietal and floral characters still fall victim to the heat. The process inevitably alters some of the honey's complex profiles and beneficial components.

- **Contamination Risks:** Though reduced, the cool-down period leaves the must vulnerable to potential airborne contaminants. However, covering the must during this phase can significantly mitigate the risk.

The beauty of Pasteurization lies in its flexibility. The must is heated to a carefully controlled temperature, less than boiling, and maintained there for a specified duration. The cooler the temperature, the longer the heating period needs to be. This approach allows nuanced control over the process, ensuring the must is sanitized without sacrificing too much of the honey's inherent goodness.

No-Boil Method

With the No-Boil method, we uncover a process grounded in simplicity and a deep understanding of honey's innate properties. Honey, in its undiluted form, is a stronghold against yeast invasion; its high sugar concentration creating a bastion where only the hardiest of yeasts might dare to venture. When diluted with water for mead-making, this defense softens, but introducing selected yeast strains ensures a controlled fermentation, pushing aside any rogue elements. This method allows the mead-maker to forge a high-quality beverage without boiling.

Why Choose No-Boil?

- **Preserving Nature's Palette:** Skipping the boil means the volatile compounds that imbue honey with its distinctive floral notes remain untouched. The result? A mead that truly sings with the essence of its honey, offering a taste and aroma profile that's as vivid as a stroll through the flowers from which it came.

- **Ease and Efficiency:** The No-Boil method is about cutting straight to the heart of mead-making without the detours of heating and cooling. It's a streamlined path, ideal for both newcomers to the craft and seasoned artisans looking for purity in their process.

- **Ready for Yeast:** With no heat to drive off, the must remains at a welcoming temperature for yeast, allowing you to move into the fermentation stage without delay.

However, the No-Boil method is not without its challenges:

- **A Closer Look at Contamination:** While honey brings its defenses, the water does not. The quality of your water becomes paramount, necessitating a source you trust, or removing chlorine to protect the delicate balance of fermentation.

- **The Stirring Saga:** Dissolving honey in water without heat will test patience and persistence! A thorough blend is crucial to avoid stratification, leading to inconsistent fermentation and unexpected flavors.

- **Water Wisdom:** If your tap water carries the scent of the swimming pool, it's time to let it sit, or opt for distilled water, ensuring nothing stands in the way of your yeast's hard work.

From my vantage point, the No-Boil method is more than just a viable alternative; it's a preferred route for those seeking to capture honey's raw, unadulterated essence. Its simplicity is not a shortcut, but a choice to preserve and respect the complexity of our primary ingredient—honey.

Step-by-Step Process

As previously mentioned, the choice of vessel is more important than you might first think. A food-grade bucket or a similar container, offering 25–50% more space than your planned batch size, is ideal. This extra room isn't just for show; it's crucial for managing the foam created during fermentation. Overflowing is a real risk, especially in containers like carboys, leading to messy and sticky situations. So, think big and give your mead the room to breathe... and froth.

Always aim a little higher with your quantities when dreaming up your mead masterpiece. For a 5-gallon goal in secondary fermentation, mix up about 5.5 to 6 gallons initially. This extra bit isn't just for sipping; it's your insurance policy. It ensures you won't fall short of your volume target despite the inevitable losses during transfer and racking. Plus, it gives you a little extra to top off the secondary vessel, keeping everything full and happy.

Pour the Honey in the Fermenting Bucket

Onto the star of the show: honey. It's sticky, it's sweet, and it's notoriously tricky to handle. Pouring honey is a lesson in patience, as it clings to everything it touches. But don't sweat the small stuff—or, in this case—the sticky strands. The aim isn't to nail the exact honey volume down to the last drop, but to get close enough to your target, without losing your sanity.

- This part of the process calls for a bit of prep and precision. With a sanitized spatula, you'll ensure no drop of honey goes to waste. Precision is key here; setting your scale to zero with the fermenter can add just the right amount of honey specified in your recipe. If you find yourself wrestling with the last bits of honey clinging to the container, a clever trick is at your disposal. A splash of the water you treated with a Campden tablet for purity can coax those final drops. Seal and shake the container if you can, or stir if it's open before uniting this honey-water mix with the rest in the fermenter. Watching the levels on your fermenter will guide you in adjusting the subsequent water addition, ensuring the balance is just right.

- After adding the honey, gently introduce warm water, approximately 40% of your batch's total water requirement, to the fermenter. This step initiates the crucial process of dissolving the honey, ensuring none remains at the fermenter's bottom. A thorough mix at this stage guarantees a smooth, evenly blended base for your mead.

- Following this, it's time to incorporate yeast nutrients into the mix. These vital components are key to energizing the yeast, and promoting vigorous fermentation. We'll explore how to effectively add these nutrients in a moment, highlighting their importance for a robust and fruitful fermentation process.

- Add the rest of the water once your must is fortified with nutrients. This is to perfect the honey-water balance, optimizing both taste and the conditions for fermentation.

- Then, the actual mixing starts with every ingredient in the fermenter. This is an essential step to ensure the complete integration of honey into the water, aiming for a perfectly consistent must. While a solid spoon is a good start, a wine degasser connected to an electric drill might be your best bet for efficiency, particularly with crystallized honey. The point to make here is to mix thoroughly but gently, gradually increasing speed to avoid spills and achieve an even mixture. Oxygenation is the final, critical step before fermentation can begin. This process involves infusing your must with oxygen, which yeast needs to thrive and multiply in the early stages. Whether using a degasser for efficiency or testing your arm strength with vigorous spoon stirring, this step ensures your yeast has the best environment for a strong, healthy start.

Mixing the Must with a Wine Degasser

Handling Crystallized Honey

Encountering crystallized honey is not the end of the world; it's merely a small bump on the road to crafting your mead. To soften it, begin with a simple, gentle method: place your honey container in warm water. This approach eases the honey towards liquidity, though it might not wholly soften very solid honey.

Turning to your oven can be a game-changer for more stubbornly crystallised honey. Setting your oven to a low and steady heat—between 125°F– 150°F (about 52°C to 66°C)—and letting the honey warm for a few hours will do wonders. This method ensures a more consistent and thorough softening, especially useful for larger quantities of honey, or when the crystals are particularly tenacious.

As you employ heat to deal with crystallized honey, keep a vigilant eye on the temperature. The goal is to liquefy the honey enough to blend smoothly into your must, without applying so much heat that you risk diminishing its rich flavors and delicate aromas. Gentle and patient warming is key, allowing the honey to release its bonds, without losing its essence.

Determining Honey Quantities for Mead-Making

When crafting mead, one of the most crucial decisions you'll face is deciding *how much* honey to use. It's a choice that doesn't come with a one-size-fits-all answer but rather depends on a blend of factors that contribute to the unique profile of each batch. Whether you're considering the volume of mead you intend to produce, the taste you're aiming for, the alcohol content you desire, or the type of yeast you plan to use—each element plays an important role in shaping the final product.

Recipes may guide you with suggested quantities, yet they often leave room for personal interpretation and adjustment. This flexibility allows you to tailor the sweetness and body of your mead according to your vision; blending art with science, in a dance of creativity and precision. In the following chapters, after having covered some key concepts, I will present you with a couple of methods that you can use to determine the honey quantity for your recipe.

Honey is the cornerstone in the alchemy of mead-making. Each pound of honey per gallon of must not only contributes to the potential alcohol content—increasing it by about 5%—but also affects the must's original gravity, raising it by approximately 35 points from water's baseline of 1.000 to 1.035. This relationship between honey, gravity, and alcohol content offers a good guide for predicting the particular characteristics of your mead.

Illustrative Examples

Here are two scenarios that highlight how honey quantity and yeast selection influence your mead:

- **Example 1:** For a batch aiming for 5 gallons, using 15 pounds of honey equates to a potential ABV of around 15%. Opting for yeast with a 14% alcohol tolerance suggests a slight residual sweetness post-fermentation, yielding a mead that leans towards dry to semi-sweet with a final gravity near 1.007. This outcome is due to the yeast's alcohol tolerance slightly under the must's potential, leaving behind a touch of unfermented sugar.

- **Example 2:** Using the same ratio of honey but switching to yeast with an 18% alcohol tolerance changes the game. This yeast can fully ferment the available sugars, likely resulting in a completely dry mead with no residual sweetness, aiming for a final gravity close to 1.000.

These examples underscore the significance of honey quantity in your mead's fermentation process, demonstrating how it influences both the ABV and the sweetness level. The choice of yeast further affects the outcome, allowing for a custom-tailored brewing experience.

Mastering Gravity: The Key to Mead's Alcohol Content

Taking a Gravity Measurement/Credit: LuYago (www .Shutterstock.com)

Mead-making brings its share of technical challenges, with the measurement of gravity topping the list of essential skills. It's a process that marries the scientific with the artisanal, allowing you to accurately predict the potential alcohol content of your mead. The hydrometer, a tool designed to measure the specific gravity, or density, of your liquid compared to water, becomes your best ally in this specific endeavor.

Taking an initial reading of your must's specific gravity is necessary at the outset of your mead-making venture. This reading, obtained by gently placing a hydrometer in your mixture, reflects the sugar content present. A scale usually spans from 1.000 to 1.160, where a higher value signals a richer sugar base, suggesting a stronger brew in the making. This initial snapshot of the density of your must sets the stage for what's to come—guiding your path in fermentation.

Turning these gravity readings into a tangible estimate of alcohol by volume (ABV) involves a simple yet enlightening formula:

$$ABV\% = (SG - FG) * 131.25$$

In this equation, SG represents the starting specific gravity, while FG represents the final gravity measured once fermentation ceases. This calculation doesn't just forecast the strength of your mead; it informs decisions for enhancing flavor and stability in the later stages of mead crafting.

Some mead artisans prefer using the Brix scale, which measures sugar content in a similar vein but requires conversion for ABV calculation:

To convert Brix to Specific Gravity:

$$SG = \left(\frac{Brix}{258.6 - \left(\left(\frac{Brix}{258.2}\right) * 227.1 \right)} \right) + 1$$

To convert Specific Gravity to Brix:

$$Brix = (((182.4601 * SG - 775.6821) * SG + 1262.7794) * SG - 669.5622$$

Despite Brix offering an alternative perspective, the directness and ease of specific gravity measurements often make them the go-to choice for home brewers. I personally recommend you consider specific gravity rather than the Brix scale.

Understanding the Fermentation Process Through Gravity

As your mead concludes its fermentation journey, capturing the final gravity (FG) reading is your next move. This action, employing the trusty hydrometer, not only sheds light on the total alcohol by volume (ABV) but also tells the story of your mead's transformation. Understanding this shift from the initial specific gravity (SG) to the final gravity offers you a full view of the fermentation's efficacy and the mead's ultimate potency.

Depending on the desired outcome, final gravity readings can vary significantly:

Dry Meads: These meads land in the range of 0.990 to 1.006, offering a crisp, clean finish.

Medium Meads: They find their sweet spot between 1.006 and 1.015, balancing sweetness with a touch of dryness.

Sweet Meads: Characterized by final gravities from 1.012 to 1.020, they embrace the luscious side of mead.

Dessert Meads: Surpassing 1.020, these are the epitome of indulgence, rich and deeply flavorful.

Why Exact ABV Matters More Than You Might Think

I understand some of you might think, "Why fuss over the alcohol content? I'm just in it for the enjoyment of brewing." And while the joy of crafting your brew is undoubtedly a big part of the process, knowing the alcohol level of your mead can spare you from potential brewing pitfalls down the line. Without tools like a hydrometer to measure specific gravity, it's virtually a guessing game to determine whether fermentation has fully completed. With its whimsical nature, yeast can pause its hard work, leading to a temporary halt in fermentation. This pause can be misleading, as fermentation might kick back into gear unexpectedly.

Imagine bottling your mead, thinking it's ready, only to find fermentation resumes! This scenario is not just disappointing; it's downright risky, posing the danger of overpressurization in the bottles. Utilizing a hydrometer to confirm that fermentation has indeed concluded, or to catch a premature pause, can help you sidestep such brewing mishaps.

What to do in Practice

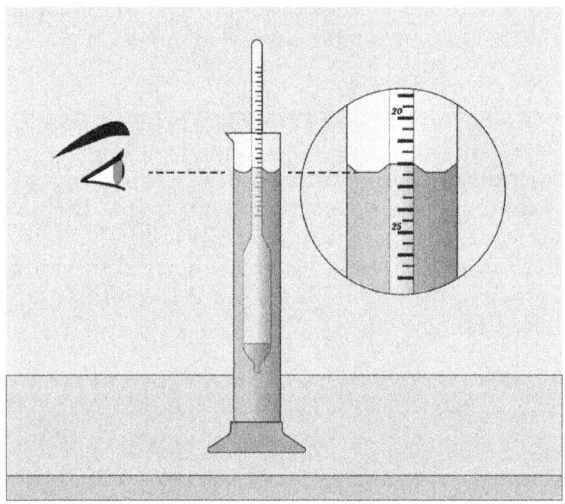

How to read an Hydromter

Having covered the reasons for incorporating a hydrometer into your brewing toolkit, let's move forward with how to apply this knowledge in practice. Ensure your hydrometer and its test jar are appropriately sanitized, (as we have seen, a crucial step to avoid introducing any unwelcome microorganisms into your mead). If you have one on hand, a turkey baster can be an excellent tool for transferring a sample of your must into the test jar without disturbing the rest of the batch. Alternatively, you can gently place the hydrometer directly into the fermenter for a more straightforward approach. This process allows you to accurately measure your brew's Starting Gravity (SG), laying the groundwork for a successful fermentation process.

Step 3: Conducting a Temperature Check

Monitoring your honey-water mixture's temperature is crucial as you navigate the mead-making process. Begin by gently inserting the thermometer into the mixture, ensuring it's fully immersed. Patiently wait for the reading to stabilize to ensure accuracy; once it has, diligently record this temperature in your brewing log. This documentation practice is invaluable for tracking your progress and refining future batches.

After noting the temperature, carefully remove the thermometer. Proceed by rinsing it in water treated with a Campden tablet. This step is vital for maintaining the purity of your brew by removing any potential contaminants. Following this rinse, place the thermometer back into the sanitizer solution to ensure it remains sterile for future use.

Finally, with the temperature accurately recorded and your equipment sanitized, secure the sanitized cover over the fermenter.

This marks the completion of the temperature check and underscores your commitment to creating a meticulously crafted mead, safeguarding the integrity of your brew at every step of the process.

Step 4: Pitching the Yeast

Assuming you've opted for dry yeast, given its consistency and user-friendly nature, let's walk through the process of introducing yeast into your must.

- **Warming the Water:** Measure the water needed for the Go-Ferm nutrient and the yeast, as specified in their packages. Warm the water slightly in the microwave; you aim for warmth, not boiling. Using a sanitized thermometer, carefully bring the water temperature in your measuring cup to exactly 104°F (40°C) by adding some room temperature water as needed. Not exceeding this temperature is crucial to ensure the yeast's viability.

- **Dissolving Go-Ferm:** Introduce the Go-Ferm into the warmed water. Use a clean spoon for stirring, aiming for complete dissolution. Go-Ferm might initially clump together but don't worry. Persistent stirring and occasionally pressing the lumps against the side of the glass will help dissolve them thoroughly.

- **Hydrating the Yeast:** Sprinkle the yeast into the nutrient-enriched water, stirring gently to ensure it's well integrated. Cover the glass with cling wrap to maintain a clean environment, and set a timer for 15 to 20 minutes. This hydration period is critical for "waking" the yeast, preparing it for its central role in fermentation.

Hydrating the Yeast

- **Introducing Yeast to the Must:** After the yeast has rehydrated, give the mixture one last stir to ensure it's uniformly suspended, then pour everything into your fermenter, which should already contain your honey-water mix. Stir the mixture gently to distribute the yeast evenly throughout the must, then cover the fermenter securely. This marks the beginning of fermentation and protects your creation from external contaminants.

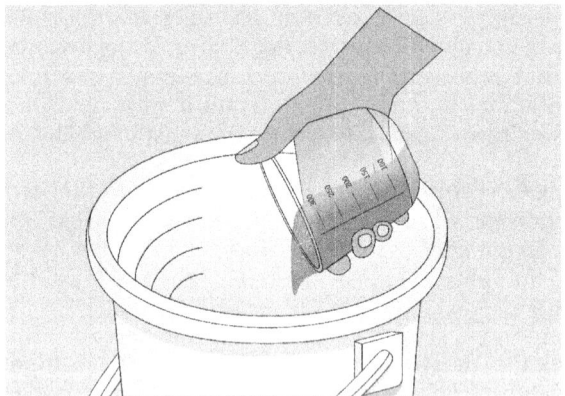

Introducing the Yeast in the Must

Step 5: Initial (Primary) Fermentation

Some mead-making terms seem to blend, creating confusion, especially when discussing "primary and secondary fermentation." Let's set the record straight and demystify what these terms mean, especially in mead-making, where we need to know the distinction between primary (or initial) and secondary (or finishing) fermentation.

Imagine the initial fermentation as the kickoff party for your mead. Here, the magic starts in the welcoming confines of the food-grade plastic vessel. This container, chosen for its practicality (easy to stir, easy to manage, and less prone to breakage than glass), is where your mead begins to find its voice. In this phase, yeast is the life of the party, feasting on sugars, creating alcohol and carbon dioxide in a kind of frenzied dance! But, like any good party, keeping an eye on the gatecrashers is necessary. Oxygen, initially consumed eagerly by our yeasty friends, can cause a party foul as fermentation slows. The trick is to move the party (your mead) to a more exclusive venue (the secondary fermenter) before oxygen overstays its welcome.

Now, think of secondary fermentation as the after-party. It's more refined, exclusive, and where the real magic happens. The venue of choice is a carboy, selected to fit just right, minimizing the space to prevent unwanted oxygen from mingling with your mead. This stage is about patience, allowing your mead to mature, clarify, and develop those complex flavors and aromas that define it. As the yeast settles down, a process known as flocculation, it's a sign that the active phase of fermentation is wrapping up. You'll know precisely when fermentation has finished by checking the specific gravity with a sanitized wine thief.

The passage of mead from initial to secondary fermentation is not just a logistical move; it's a transformation. Initial fermentation is vibrant, bustling with activity as yeast transforms honey into something magical. It's all about setting the stage and creating the right environment for fermentation to kick off with gusto. Secondary fermentation, on the other hand, is the maturing phase, where the mead gets to refine its character, clarity, and flavor in a more controlled and serene setting.

Understanding Primary Fermentation

Your must is now nestled in a cool, dim spot, ideally between 65–70°F, perfectly poised for the magic to begin. It's a waiting game at this stage, with the anticipation building a few hours after you've introduced the yeast into the mix. Patience is key here; the initial quiet is deceptive, signaling the yeast's critical acclimatization phase, known as the "lag phase", rather than mere inactivity. During this time, the yeast cells adjust to their new environment, gearing up for the task ahead by absorbing essential nutrients. This adjustment period varies, influenced by factors such as the must's specific gravity, acidity, ambient temperature, the initial yeast population, and available nutrients, typically ranging from a few hours to a bit longer.

To encapsulate the yeast's journey briefly:

- **Lag Phase:** This is the yeast's orientation period in the must, where they prepare for the feast by taking in nutrients necessary for growth and repro-duction. The end of this phase is heralded by the appearance of "krausen", a frothy layer that forms on the surface of the must, indicating that it might be time to introduce additional nutrients to bolster the yeast's activity.

- **Growth Phase:** Following the quiet of the lag phase, the yeast spring into action, multiplying exponentially as they feast on the available nutrients. This growth spurt is when your job in aeration becomes crucial, helping to stir the pot and ensuring the yeast has access to the oxygen needed for this vigorous expansion.

- **Fermentation Phase:** The scene shifts as the yeast transitions from an oxygen-reliant (aerobic) to an oxygen-independent (anaerobic) metabolism. This switch allows them to consume the sugars in the honey, transforming them into ethanol, the alchemical gold of mead-making.

The Vitality of Lag Phase and the Wisdom of Staggered Nutrient Addition

Stepping back into our mead-making saga, let's recall that honey, our star ingredient, is quite an ascetic with regards to nutrition. It offers a minimalistic array of free amino nitrogen (FAN) and other trace elements, which our microscopic yeast workers need to thrive. To bolster their efforts, we turn to a duo of nutritional reinforcements: diammonium phosphate (DAP) and Fermaid K, *or* just to Fermaid O.

In the delicate early days of fermentation, staggered nutrient additions are like the timely checkpoints in a marathon, keeping the yeast energized and on pace. A rule of thumb for a standard 5-gallon batch is to use 4 grams of Fermaid K and 8 grams of DAP. You can measure these nutrients into a sealable bag at the onset, placing it beside the fermenter. It's a simple trick to ensure consistency, without getting lost in the minutiae.

You can follow this schedule:

- Introduce 2 grams, roughly equivalent to 3/4 teaspoon, after the first 24 hours have elapsed.

- Continue the process with an additional 2 grams after the first 48 hours have elapsed.

- Proceed with another 2 grams once you reach the 72-hour mark.

- Conclude your nutrient additions with 2 grams when the specific gravity (SG) indicates that the sugar content has decreased by 30%, or on Day 7.

The nutrient additions are carefully scheduled to align with the yeast's performance. Picture the yeast as a troupe of performers, with the first three acts of nutrients supporting their exponential growth phase. The final act, delivered at the point when 30% of the sugar has been converted, serves as an encore, providing a last boost as the yeast settles into its stationary phase.

On Day 4, it's time to take the stage with your hydrometer and check if the must has hit the 1/3 sugar break—a clear sign of progress. The SG should have reduced by 1/3 versus the initial value. If it's a go, add the final portion of nutrients and gently stir the mead. If not, you'll continue daily checks, ready to act when the moment arrives.

Navigating Nutrient Additions with Care

As you reach the important moments of adding nutrients in your mead-making process, it's essential to approach with caution. Dry powders have a knack for triggering effervescence, becoming the epicenters for carbon dioxide escape from your fermenting concoction. To sidestep the distinct possibility of creating a mini spectacle—reminiscent of a science experiment gone very awry!—it's wise to introduce these nutrients only *after* giving your mead a thorough stir. This way, you circumvent the risk of turning your workspace into a very sticky laboratory!

If you prefer a method that's even more controlled, consider premixing your nutrients with a bit of water or some of the must itself in a separate jar. By doing this, you can add this nutrient solution to the fermenter with finesse, drastically diminishing the chances of any volcanic foam-over.

Fermaid O

For those opting for Fermaid O in their mead venture, a conversion formula is at your disposal to fine-tune the quantities:

To substitute for 1 gram of Fermaid O, use 0.40 grams of Fermaid K or 0.19 grams of DAP.

Furthermore, keep a close eye on the Yeast Assimilable Nitrogen (YAN) levels to tailor your nutrient additions accurately:

- With Fermaid-O, aim for 40 parts per million nitrogen per liter, capping it at 450 mg/liter to maintain the mead's flavor profile without overpowering yeasty notes.

- Fermaid-K, which provides 100 ppm N/L, should be used up to 500 mg/liter, aligning with commercial standards set by the TTB, primarily due to the thiamin content.

- Diammonium Phosphate offers a robust 210 ppm N/L, but the TTB caps its use at 960 mg/liter. Remember, as fermentation progresses and reaches around 9% ABV, the yeast's ability to utilize DAP wanes due to the alcohol's presence.

- Lastly, Go-Ferm provides a more modest 30 ppm N/L, specifically reserved for the initial rehydration stage, utilized within the first 24 hours.

For those eager to learn more about the precise calculation of nutrient requirements for your mead, a dedicated chapter appears later in this book. I suggest you familiarize yourself with the nutrient quantities I've outlined. As you gain more experience and confidence in your mead-making, you'll find yourself refining your approach to managing YAN requirements and planning nutrient additions more adeptly.

If your curiosity gets the better of you, feel free to leap ahead to the "Advanced Nutrient Addition" chapter. However, I would recommend starting from the beginning of this book, particularly if you're new to the craft. This structured approach will build a solid foundation, ensuring you're well-prepared to tackle more complex topics with ease.

Aeration

Oxygen, often termed "the elixir of life", is just as vital in our subject, the realm of mead, as it is in our atmosphere. Yeast cells depend on it to perform their fermentation duties effectively. Without oxygen, they can't fully engage in the transformative process that converts honey into mead, leaving behind carbon dioxide as evidence of their hard work, visible through the rhythmic dance of bubbles escaping the airlock.

The initial days of fermentation are a delicate growth phase for yeast, akin to the early development of a seedling. Just as a young plant needs ample sunlight, yeast requires a generous oxygen supply. A simple yet effective way to provide this is to aerate the must. This can be done several times a day, during the first three days, by employing a wine degasser or engaging in the age-old practice of shaking the container. For those who prefer a more high-tech approach, an oxygen tank paired with an oxygen stone can deliver a pure, potent dose directly into the heart of the must, which the yeast will readily consume.

Let´s face it, not everyone has an oxygen tank—but that shouldn't be a barrier. Aeration tools that pair an aerator with a stone can be obtained at modest cost, offering a more accessible means to the same end.

Before introducing air or oxygen into your must, it's imperative to sanitize all equipment. As we have seen, this preemptive strike against contaminants is a non-negotiable part of responsible mead-making. Once your tools are sanitized, aerate the must for two to five minutes, depending on your chosen method. This simple act can profoundly influence the vitality of your fermentation.

This process has an important caveat: carbon dioxide dissolved in the must can cause "a bit of a spectacle" if not handled with care! A sudden introduction of oxygen could trigger a geyser effect, turning your mead-making session into a cleaning ordeal.

To avoid this drama, gently stir the must to release the trapped gases before proceeding with oxygenation. This step is *especially* crucial if shaking is your preferred method of aeration.

Fine-Tuning pH for Optimal Fermentation in Mead-Making

The mead´s pH levels act as critical waypoints for successful fermentation. The must's natural acidity will increase during fermentation, affecting the pH. The sweet spot for most mead fermentations is nestled within the pH range of 3.4 to 4.0, creating the perfect environment for yeast to work their magic.

Before beginning fermentation, it is wise to adjust your must's pH to about 4.0. This foresight can prevent the potential stall of fermentation due to a significant drop in pH. Potassium carbonate emerges as the hero in this scenario, not just for its pH-adjusting capabilities but also for enriching the must with potassium, which is essential for robust yeast health.

Tread lightly with potassium carbonate. An overzealous addition could diminish the must's total acidity, disrupting the delicate balance of the final mead and possibly introducing unwanted flavors. A guideline to follow is to restrict potassium carbonate usage to no more than 5 grams in a five-gallon mead batch for pH adjustments.

A high-quality, calibrated pH meter is your best tool for measuring pH. While pH test strips can offer a general idea, they may fall short of the accuracy needed for fine adjustments. Always test your must before fermentation to ensure it's not starting from a low-pH disadvantage. Drawing a small sample for testing, rather than dipping instruments directly into the fermenter, preserves the purity of your must.

Potassium's contribution to buffering the must cannot be overstated, with optimal levels reaching above 300 ppm to maintain appropriate pH. Introducing 5 grams of food-grade potassium carbonate can add roughly 136 ppm of potassium, gently raising the pH to more yeast-friendly levels.

Should obtaining potassium carbonate prove elusive, calcium carbonate, or chalk, offers a readily available alternative. Available at most homebrew shops, it provides a reliable means to adjust pH when potassium carbonate isn't an option.

As your mead ferments, keep a close eye on its progress. If you notice any unexpected slowdowns within the initial fermentation phase—it's time to test the pH again. This continual monitoring ensures that your mead stays on the right track toward becoming a well-rounded, flavorful delight!

Degassing

Degassing occupies a crucial place in the craft of mead-making, addressing the challenge of carbon dioxide (CO_2) buildup during the primary fermentation phase. As the yeast works its magic, transforming available oxygen in the mead into CO_2, not all this gas finds its way out through the airlock, leading to its retention within the mead.

Distinguishing Between Degassing and Oxygenation

It's essential to differentiate between the processes of degassing and oxygenation within mead-making. Oxygenation is beneficial at the *beginning* of the fermentation process, supplying the yeast with the oxygen necessary for a robust start. However, once fermentation concludes, oxygen becomes undesirable, as any lingering yeast may consume it and continue producing CO_2, potentially altering the mead's character.

Effective Techniques for Degassing

To efficiently remove excess CO_2 from your mead, gentle stirring or swirling within the container is recommended. This can be achieved with a basic spoon for smaller batches or a wine degasser for larger volumes, significantly easing the task. The emergence of bubbles on the container's sides or within the mead itself is a positive sign, indicating the release of CO_2 and the effectiveness of your degassing efforts.

Utilizing a "Wine Saver," a device commonly used for preserving wine, presents another method for degassing. This tool creates a vacuum, extracting air and CO_2 from the mead, a process visually confirmed by the appearance of bubbles. This technique is particularly efficient.

Overlooking the degassing process can lead to complications during bottling and long-term storage. Excessive CO_2 can cause the mead to become over-carbonated, risking bubbling or exploding when eventually opened. Keeping airlocks on the mead containers as long as feasible can act as a passive degassing method, facilitating the gradual escape of CO_2.

The frequency of degassing is a subjective process and varies with each batch of mead. Observing the mead for signs of bubbling during the degassing process provides a practical measure of CO_2 levels. Reduced bubbling suggests successful degassing, while persistent bubbling may indicate more CO_2 needs to be released. Post-degassing, allowing the mead to settle for 8 to 12 weeks is advantageous. This period lets the mead stabilize, enabling it to recover the complexity and depth that might have been diminished due to degassing.

Step-by-Step Guide to Fermentation Management

With this trove of fermentation knowledge at your fingertips, let's distill all this into actionable steps!

Your first **eight-day** mission is to **stir** the fermenting honey **two to three times daily**. Ideally, you'd space these stirrings out every **8 to 12 hours**. Life can be unpredictable, so strive for consistency, but don't fret if your schedule isn't clockwork perfect. When you stir, start slowly. You'll notice a frothy buildup—this is normal. Continue stirring until the foam ceases to form. It's a simple yet vital routine that ensures an even fermentation and prevents any off-flavors from developing.

Each stirring session is an entry in the story of your mead. So, **jot down** every detail in your fermentation log—the date, time, and what you observe. Before you begin stirring, place a sanitized **hydrometer** in the fermenter to get a real-time snapshot of your mead's progress for your records.

As you navigate through **days 2, 3, and 7**, remember these are your checkpoints to bolster the must with **nutrients**. After your daily stirrings, introduce another 2 grams of the Fermaid K and diammonium phosphate blend. This scheduled reinforcement is designed to sustain the yeast's momentum. It's like giving them a second wind, ensuring they have the resources needed for the continuing work ahead.

Continue this **stirring** ritual diligently, but lay down your stirring spoon once you reach **day 8.** From this point on, it's about **observation.** You're watching for the telltale signs that the primary fermentation is reaching its grand finale.

The end of fermentation isn't a moment but a series of consistent **hydrometer** readings over several days. When these numbers stabilize, and the anticipated decrease in sugar content halts, you'll know the primary fermentation has completed its course.

Observing the airlock's bubbling frequency is another practical method to gauge this activity. This doesn't require constant vigilance but, instead, a **check-in** every few days to assess the vigor of the fermentation process. When you notice that the bubbling pace has decelerated to about **one bubble every 30 seconds,** the fermentation has reached a pivotal moment. It signals that it's time to transition your mead to the next crucial phase, **racking.**

Preventing and Handling Overflow

Mead Overflow/Credit: BDoss928 (www.Shutterstoc k.com)

Even the most practised flights can encounter turbulence, like when you find your mead eagerly trying to escape its vessel during primary fermentation! Don't fret! Here's how to keep the magic *inside* the cauldron:

- **Opt for Roomier Quarters.** Imagine giving your mead the space to frolic and foam without worrying about an impending mess. That's exactly what you're doing when you choose a fermentation vessel that's just a bit roomier than your batch size. If you're working on a gallon of mead, a 2-gallon food-safe plastic bucket becomes your best friend. And for those ambitious 5-gallon batches, a 7.9-gallon bucket is your ally. These buckets are budget-friendly and a breeze to find at your local homebrew shop.

- **The Blowoff Tube:** A Trusty Sidekick. A blowoff tube can be your mead's knight in shining armor if you're in a pinch, and a bigger bucket isn't on the cards. It's a simple setup: a tube running from the fermenter's lid down into a container of sanitizer. This acts like an oversized airlock, giving the foam a path to freedom without the mess. Ensure the tube's end stays submerged in the sanitizer to keep things clean and sealed. Yet, let's get real—while a blowoff tube can save the day, it's more of a band-aid solution. Vigorous fermentations can *still* clog up the works, especially those with fruits or other additives. If you go down this route, keep a watchful eye on it. You might have to play the hero and unclog it... to prevent an overflow disaster.

Step 6: Racking; Siphoning & Finishing (Secondary) Fermentation

Secondary fermentation is where the rhythm slows, allowing the mead to mature and develop its full character. This stage is traditionally conducted in a carboy, often glass, setting the stage for quiet and calm after the lively primary fermentation. Here, the focus shifts from vigorous fermentation to finessing the mead towards clarity and stability, readying it for its final presentation.

The transition from the primary fermenter to the carboy is critical in the mead's journey. Before this move, ensuring that every piece of equipment, especially the carboy, is scrupulously clean and sanitized is paramount. The mead is then transferred—or racked—using a siphon, separating it from the sediment that has built up, a testament to the yeast's hard work during the initial fermentation.

Headspace: A Crucial Consideration

An often-overlooked aspect of this phase is managing the headspace (air layer between mead and lid) within the carboy. The goal is to minimize this air gap to protect the mead from oxidation, which can tarnish its flavors and aromas as it matures. By choosing a carboy that snugly fits the volume of your batch, you significantly reduce oxygen exposure. For larger batches that necessitate splitting the mead into several carboys, thoughtful distribution can help maintain the integrity of the mead. A novel approach to reducing headspace involves the use of sanitized glass marbles. You can displace the mead by carefully adding it to the carboy, reducing the air volume above it. This technique requires a gentle hand to prevent damage to the glass carboy.

Ensuring Fermentation's End

As the mead settles, the yeast flocculates (forms large flocs and then drops out of suspension), signaling the fermentation is nearing its end. However, not all yeast strains settle out with the same efficiency, making visual cues less reliable. A systematic approach involves taking specific gravity readings periodically with a sanitized instrument. Consistent readings over consecutive days confirm that fermentation has ceased, an essential detail to note in your brewing log. Let's discuss the processes of racking and siphoning:

Racking

This step involves carefully transferring mead from one vessel to another, aiming to achieve several key purposes. The primary aim of racking is to bring the fermentation process to a definitive halt. This is accomplished by separating the mead from the active yeast sediment.

A significant concern in mead-making is the potential for off-flavors to arise from prolonged contact with the lees—the layers of sediment composed mainly of yeast cells that accumulate at the bottom of the fermenter. Racking allows for removing mead from these sediments, thereby avoiding the yeasty flavors resulting from the breakdown of yeast cells, especially critical before the mead undergoes bulk aging.

Although not applicable to *every* batch of mead, racking offers a perfect opportunity to introduce additional ingredients that can infuse the mead with complex flavors during secondary fermentation. This aspect of racking allows for creative experimentation with various flavor profiles, though it's noted that our current mead preparation does not involve this step.

Another key aspect of racking is its role in clarification. Transferring the mead away from the lees significantly enhances its clarity, preparing it for a visually appealing presentation upon bottling.

A crucial consideration during racking is the careful management of oxygen exposure. Excessive air mixed into the mead during this process will introduce oxygen, potentially negatively affecting the mead's flavor. Ensuring all equipment is thoroughly sanitized, and handling the mead gently can mitigate these risks.

It's essential to recognize that racking, while beneficial for clarification, flavor enhancement, or back-sweetening, is not an *absolute* requirement before bottling. You might keep the mead in the fermenter until the bottling stage. However, this approach risks the transfer of yeast, which could impact clarity and taste of your brew.

Siphoning

If not very carefully executed, this next step holds the potential for issues, including the risk of contamination or unwanted aeration of your mead. Begin by positioning your mead on a table at least one day before siphoning. This preparatory step allows the yeast to settle, reducing the likelihood of its transfer. It's also essential to keep the siphon's end submerged in the mead to prevent air exposure, which could jeopardize its integrity.

There are three primary methods for siphoning, each with its own set of considerations:

125

- **The Sucking Method:** While straightforward, this technique poses a contamination risk through contact with the human mouth. To mitigate this, one could rinse their mouth with a strong alcoholic solution like vodka (why not!) or peppermint schnapps, though this method still carries inherent risks.

- **Water Priming:** This method is both simple and effective. Begin by placing the secondary carboy within a larger, empty vessel. Fill the racking cane and hose with sanitized water—avoiding rinse-required solutions—and seal the ends with clean fingers to retain the water. Position the hose's exit below the mead's surface level, insert the racking cane into the mead, and release the exit end to initiate flow. Once the mead starts to flow, quickly reseal the end, transfer it to your carboy or bottles, and then allow the mead to flow freely.

- **Using an AutoSiphon:** This handy tool simplifies the siphoning process, enabling flow initiation through a simple pump action. It's easy to use and maintains sanitation, though it might not fit into all container types, such as the necks of 1-gallon wine jugs. This is my preferred method and the one I'll detail further.

Of all these methods, the AutoSiphon stands out for its efficiency and hygienic advantages, making it the recommended choice for transferring mead. Its design allows for a straightforward operation, significantly reducing the risk of contamination and ensuring a smooth transfer of your mead into secondary containers or bottles.

Step-by-Step Process

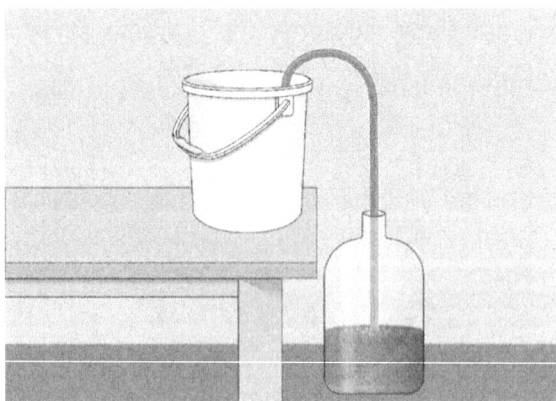

Racking Mead

Here is a detailed, step-by-step walkthrough to provide clarity and precision in this critical process.

- **Setting Up for Siphoning:** Begin by elevating your fermenting bucket onto a raised surface, such as a table or countertop, to facilitate the siphoning process into the carboy. First, sanitize your carboy thoroughly by swirling

126

around a liter of sanitizer solution, ensuring complete coverage inside. After sanitizing, empty the remaining solution into your sanitizer bucket and position the carboy on the floor beneath the mead bucket.

- **Sanitizing Equipment:** Next, it's time to sanitize all the tools involved in the transfer – this includes the carboy stopper, airlock, auto-siphon, and racking hose. Immerse the auto-siphon and racking hose in the sanitizer solution, ensuring they are thoroughly soaked, then allow them to drain. Assemble your siphoning setup by inserting the auto-siphon into the mead bucket and attaching the racking hose.

- **Initiating the Transfer:** Position a hydrometer sample jar near the carboy on the floor. Begin the siphoning process using the auto-siphons pump, directing the initial mead flow into the sample jar. Once nearly full, seal the hose with a sanitized finger, then carefully introduce the hose into the carboy, aiming for the bottom to minimize splashing. Continue the transfer until complete, then seal the carboy with the stopper and airlock. Document the racking date and time in your brewing log. Then, measure the mead's specific gravity with the hydrometer and note the temperature using a thermometer, recording both in your log.

- **Monitoring Secondary Fermentation:** Relocate the carboy to a stable, room-temperature area away from direct sunlight. Periodically check on the fermentation activity. You should observe a significant reduction in airlock bubbles, eventually dropping to less than one per minute, indicating the fermentation is slowing. This phase also encourages clarification as yeast cells begin to settle.

Multiple Racking

Multiple racking sessions are pivotal in enhancing the mead's quality by aiding in degasification and clarification. This technique, while beneficial, comes with its considerations, notably the potential loss of mead volume with each transfer. This loss can vary; sometimes its minimal, other times more significant—especially when dealing with batches that contain excessive pulp.

A proactive strategy to counteract this loss of volume involves preparing additional must at the outset of the brewing process. This foresight allows for the replenishment of mead in secondary carboys, ensuring that the volume remains optimal throughout the aging process. This extra step compensates for any loss incurred during racking and maintains the integrity of the mead's flavor profile.

Step 7: Bottling

Bottling/Credit: Hans Geel (www.Shutterstock.com)

As your mead's fermentation journey concludes and you've fine-tuned it with any last adjustments, the focus shifts to bottling— an important step that's as much about preservation as it is about preparation for your next batch. The choice of containers and closures can significantly influence your mead's longevity and character.

- **For Beer Bottle Enthusiasts:** If you're leaning toward beer bottles, you must secure them with crown caps. It's essential to crimp these caps tightly onto the bottles using a bottle capper, ensuring a seal that keeps the mead safe and sound.

- **Wine Bottles & Corks:** Opting for wine bottles introduces the choice between corked or screw-top closures. Corking devices vary widely, from budget options that might test your patience to more sophisticated, user-friendly models that make sealing a breeze. Traditional wine bottles, when paired with corks, are particularly suited to still meads. However, if your mead has a sparkle, it demands a more robust approach to withstand the internal pressure.

- **A Word on Sparkling Meads:** Sparkling meads require a closure system to handle the fizz. A wire-caged cork, like those used in champagne bottles, ensures the cork remains in place despite the pressure. These special corks and bottles designed for carbonation (noted by their punted bottoms) are non-negotiable for sparkling varieties.

- **Reconsidering Plastic Bottles:** Typically, plastic bottles are not favored for long-term mead storage due to potential oxygen infiltration, which can compromise the mead's quality. While some modern plastics claim to offer an oxygen barrier, their efficacy for mead has yet to be proven.

- **Enhancing Seal Integrity:** For those planning to age their mead, oxygen-barrier crown caps offer an added layer of protection thanks to a special liner that combats oxygen exposure. Alternatively, combining a cork with a crown cap can further secure your mead against oxygen ingress.

- **Presentation Matters:** For those looking to elevate their mead's presentation, exploring unique bottle shapes, colors, and sealing mechanisms can add an element of visual appeal. Swing-top bottles, for instance, offer a resealable option that marries functionality with style, available in a variety of sizes and glass colors.

Choosing the right bottle often depends on how you plan to enjoy your mead. Smaller 12 oz. beer bottles are ideal for sipping slowly, while screw-top bottles can offer convenience for those who wish to consume their mead soon after it's bottled.

Clarifying Before Bottling

A critical step before bottling is ensuring the mead's clarity. If the mead still shows signs of cloudiness a week before bottling, another racking might be necessary. The goal is to bottle *crystal clear* mead, reflecting the quality and purity of your craftsmanship.

Preventing Unwanted Fermentation

The last racking before bottling is also a moment to assess ongoing fermentation. It's good to introduce stabilizers like Metabisulfites and Sorbates to halt any remaining yeast activity, securing the mead's stability for bottling (we will cover this operation in detail when I discuss mead stabilization).

Managing the Headspace

Proper bottling technique is essential to maintain the integrity of the mead. Managing the headspace in each bottle is key to ensuring the corks remain in place. A reliable method involves filling the bottle to the top and then withdrawing the filler tube, which naturally leaves the ideal amount of headspace. This practice minimizes the risk of corks loosening, which can compromise the mead's quality over time.

Step-by-Step Guide to Bottling

Bottling your mead is a fun and rewarding phase in the mead-making process, encapsulating the essence of your hard work into each bottle. To ensure a smooth transition from carboy to bottle, here's a methodical approach to follow:

- **Setting Up for Bottling:** Position your carboy filled with clear mead on a counter or table. The height is crucial for facilitating the siphoning process effectively into the bottles.

- **Sanitization Process:** Sanitize each bottle by pouring a small amount of sanitizer solution into one, covering the top with your sanitized thumb, and shaking well to coat the interior. Empty the sanitizer back into its container and arrange the bottle upright on the floor. Repeat this process for all bottles. Also, sanitize the bottle caps, hydrometer sample jar, measuring cup, and bottling wand in your sanitizer solution, ensuring thorough coverage. After sanitizing, drain the equipment and prepare for bottling by setting up the autosiphon and racking hose for use.

- **Transferring the Mead:** Activate the autosiphon to initiate the mead's transfer, allowing it to flow into the hydrometer sample jar through the bottling wand. When the jar is almost full, position the bottling wand above the sanitized measuring cup as you prepare to fill the bottles.

- **Sealing the Bottles:** Once the bottles are filled, seal them while they are still on the counter. Using a bottle capper, securely crimp each cap onto the bottles. Ensure each bottle is sealed correctly before logging the bottling date and the total number of bottles for your records.

- **Labeling Your Bottles:** Labeling is essential for your mead's long-term identification and organization. Labelling the cases or six-packs and each bottle cap is advisable. Utilize small, round stick-on labels for a neat and efficient system. If using a coding system, avoid overly cryptic abbreviations that might be confusing in the future. Including batch numbers that refer back to a brewing journal, and the bottling date on each label will greatly aid in tracking and enjoying your mead in the coming years.

Conclusion

We've walked along the fundamental mead-making path, focusing on the essential ingredients and steps that form the backbone of traditional mead-crafting. From selecting your initial ingredients to the pivotal moment of bottling your creation, this story has been about building a solid foundation for your ongoing expertise.

Each step has been outlined with an eye for simplicity and effectiveness, ensuring a clear road to follow as you start your mead-making adventure! These initial steps will instill confidence in your practice and provide a reliable blueprint for creating very delightful mead.

More sophisticated practices are covered in the next chapter. As your familiarity with the process grows, and your skills sharpen, you'll be ready to explore various techniques to enhance your mead's complexity, flavor, and overall quality.

CHAPTER 5
Advanced Techniques And Methods

This chapter opens up the intricate world of advanced techniques and methods, elevating your mead-crafting to new heights. It's a guide for those who've mastered the basics and are now eager to explore mead's depth of flavor and complexity. We'll delve into the finer points that make an *excellent* mead *great*, focusing on stabilization, back-sweetening, the art of aging, and more. This journey is about precision, patience, and *passion*—blending science with the magic of ancient traditions. It's your invitation to push boundaries, experiment with new flavors, and create meads that are varied, rich and layered. Here, every technique is a step toward perfection, and every tip is a tool for transformation.

Mead Stabilization

Stabilizing your mead is essential for the final quality of your brew. It's the safeguard that ensures fermentation has truly finished, allowing you to bottle your mead without worrying about continued fermentation that could affect flavor or cause bottles to burst. This stability also allows you to sweeten your mead further, knowing the sweetness will remain as intended without fermentation. A stabilized mead is better armored against the forces of degradation, and the risks of infection.

When to Stabilize

Knowing the best time to stabilize can be a bit of a balancing act; it comes after fermentation has stopped and the mead has started to clear. There's no need for absolute clarity at this point—a little patience will often take care of that—but if you jump the gun, you might interrupt those final, subtle stages of fermentation where the yeast does their last bit of cleanup, which can be quite influential on the mead's final flavor profile.

It's worth noting that some meads will stabilize naturally. This is more likely with meads of normal strength, fermented to complete dryness. But just leaving your mead's stability to chance, especially if you plan to enhance the sweetness or store it for future enjoyment, is not advisable.

Before you consider your mead ready for long-term storage or bottling, it's critical to ensure it's stable. This means taking specific gravity readings after additional sweetening, and then checking again after a week. Consistent readings indicate you're good to go; changes mean more work must be done.

The following sections look at the various methods you can use to stabilize your mead.

Stabilization of Mead Using Chemical Additives

Regarding stabilization, chemical additives offer a reliable route to ensuring your mead remains unaltered after fermentation. The go-to duo for this purpose is potassium metabisulfite and potassium sorbate, each playing a distinct role in the process.

Potassium metabisulfite, commonly called "k-meta," is instrumental in scavenging oxygen from the mead, inhibiting the yeast population's ability to kickstart fermentation again. Potassium sorbate, known as "k-sorb," complements k-meta's action by hindering the remaining yeast's reproduction ability, rendering them sterile. Together, these additives form a formidable defense against any future fermentation—though it's crucial to note that they may not be able to *completely* halt an active, robust fermentation.

If you're planning to back-sweeten your mead, a prudent approach is to wait about 24 hours after adding these stabilizers before introducing additional sugars. This waiting period safeguards against the potential reactivation of fermentation, especially when racking mead that contains residual sugars.

It's essential to use potassium metabisulfite and potassium sorbate only under specific conditions. These additives are most effective when the mead has fermented to dryness, or when fermentation has ceased due to other factors, such as reaching their alcohol tolerance. To ascertain that fermentation has stopped, take hydrometer readings one week apart. Consistent readings signal that it's safe to proceed with stabilization; discrepancies indicate that fermentation is still underway, and thus, chemical stabilization may not be successful in preventing further fermentation.

While both additives are generally used in tandem, it's worth noting that potassium sorbate's effectiveness diminishes in meads with an alcohol content above 14–15% ABV. In such cases, relying on potassium metabisulfite may be the preferred approach.

Potassium Metabisulfite

Potassium metabisulfite is commonly known by various names, including PM, K-Meta, or the more household term, "Campden Tablets".

This compound is a key player in mead preservation due to its antioxidant properties. As mead ages, it's vulnerable to the detrimental effects of oxygen, which can alter color and flavor. By removing free oxygen, potassium metabisulfite not only preserves the mead's integrity but also inhibits the vitality of yeast cells, particularly those already under stress.

For homebrewers, potassium metabisulfite is accessible in two primary forms: the convenient Campden Tablets and the raw powdered version. The tablets offer a user-friendly, pre-measured dose of K-Meta, making them a practical choice for those without a precision scale. It's worth noting that one Campden Tablet typically contains .44 grams of potassium metabisulfite. However, caution is advised to avoid tablets made with sodium metabisulfite, as they can impart unwanted sodium into your mead, potentially affecting its taste. Packaging guidelines often recommend using one tablet per gallon for adequate stabilization.

In contrast, the powdered form of potassium metabisulfite affords more precise measurement, especially if you possess a gram scale that measures to the nearest tenth of a gram. It's ideal for those meticulous about their dosing, but requires a cautious approach to avoid inhaling dust.

When introduced to water, potassium metabisulfite decomposes into sulfite, bisulfite, and sulfur dioxide (SO_2), each capable of binding with oxygen molecules. This binding action protects the mead from oxidation and hinders spoilage organisms and residual yeast from utilizing the oxygen. However, the free sulfites generated can bind with other compounds or dissipate as gas, affecting the precise dosing.

When using Campden Tablets, it's best to crush them into a fine powder, ensuring they dissolve more effectively. Dissolve the required amount in a small mead sample if you're working with tablets or pure powder. Then, reintegrate this solution into the main batch, mixing gently yet thoroughly to ensure an even distribution. If you're racking simultaneously, siphoning onto the powder at the bottom of the new vessel will naturally disperse the potassium metabisulfite throughout the mead.

Potassium Sorbate

Potassium sorbate is a preservative known to many mead-makers as an essential ally in the battle against unwanted fermentation. It's referred to in the trade by its initials, PS, or its chemical monikers, K-Sorb or K-Sorbate, and it's vital for maintaining the integrity of your mead after the fermentation process has ended.

Potassium sorbate has the ability to halt any lingering yeast aspirations to multiply, ensuring the mead you've painstakingly fermented doesn't bubble back to life once bottled. It's crucial when you've reached the end of fermentation, and only a handful of yeast cells remain. The dosage isn't one-size-fits-all; it depends on the specific attributes of your batch of mead.

Adding potassium sorbate to water breaks it down into two elements: sorbic acid and potassium. Sorbic acid forms the lion's share of potassium sorbate's composition and acts as an antimicrobial agent. However, it's not a panacea; not all microbes fall under its spell. That's why it's critical to confirm your mead's stability before introducing potassium sorbate, and to always pair it with potassium metabisulfite for a comprehensive defensive strategy. It's worth noting that certain bacteria, like *Lactobacillus*, process sorbic acid into a substance reminiscent of the aroma of geranium leaves.

There's a twist: in the presence of alcohol, sorbic acid can evolve into ethyl sorbate, a compound that may impart a pineapple or celery aroma. This could introduce a ticking clock on your mead's peak condition, but these changes are typically subtle and may not even be detectable in low concentrations. This fruity nuance can sometimes add quite a desirable complexity to your mead.

The matter of dosing potassium sorbate is a topic of much debate. There's no definitive consensus, with the most cited reference being a study from the 1980s by the French enologist, Émile Peynaud. He observed that sorbic acid's efficacy is notably diminished at higher pH levels but also noted that the presence of alcohol augments its performance. His guidelines were based on a pH under 3.5, with the assumption of adequate SO_2 protection already in place.

As an adept mead-maker, you'll weigh the necessity of potassium sorbate against factors such as the mead's alcohol content, the aging process, and other stabilization methods you might employ, such as filtration or pasteurization. While potassium sorbate is an effective tool, especially for beginners or those with limited resources, it's not always required.

Be mindful of the potential long-term effects of potassium sorbate. Lactic acid bacteria can metabolize sorbic acid over time, leading to a mild off-flavor known as "geranium taint". Additionally, some mead-makers who take a more natural approach may prefer to avoid inorganic additives. Despite these considerations, potassium sorbate remains a useful stabilizer for many, particularly those new to the craft or working with mead styles less suited to the alternative stabilization techniques.

On average, a flat teaspoon of potassium sorbate will contain approximately 1.9 to 2.0 grams of the substance. When it's time to add it to your mead, patience is key. It dissolves with difficulty, so the best method is to dissolve it in a small mead sample, similar to the technique used with potassium metabisulfite. Simply adding it to an empty carboy before racking may not provide sufficient agitation for it to dissolve properly.

Harnessing Yeast Alcohol Tolerance to Stabilize the Mead

Choosing the right yeast strain and properly managing nutrient additives can lead to mead that naturally halts fermentation due to the yeast's alcohol tolerance. This approach gives a mead with residual sugars, achieving a medium-sweet to sweet profile. It's worth noting, however, that this method might yield results a few points off the manufacturer's stated value for yeast alcohol tolerance. Yet, consistency in your brewing process—particularly concerning yeast strain, sugar levels, pH, and targeted final gravity—can make this outcome repeatable.

Initiate the process with a specific gravity high enough to potentially reach the yeast's alcohol tolerance limit. Adequate nitrogen from nutrient sources should facilitate reaching this threshold. A slight shortfall from the target isn't cause for concern, as the primary goal is to naturally cease fermentation by pushing the yeast to its limit.

Keep a close eye on the specific gravity as fermentation nears its end. When readings drop below 1.010, incrementally add honey to elevate it back to between 1.015 and 1.020. For those mindful of the final alcohol by volume (ABV), record the additional honey added to adjust the gravity, as this will factor into the overall ABV calculation. Eventually, fermentation will conclude, stabilizing the gravity within the desired sweet range of 1.010–1.020. Further honey can be added to taste for an even sweeter mead, though this may temporarily reignite fermentation.

Even when leveraging alcohol tolerance as a natural stop to fermentation, the optional use of chemical stabilizers can further secure the mead against reactivation. This step requires vigilant observation to ensure fermentation does not resume.

Relying solely on ABV to arrest fermentation does carry a reactivation risk, particularly under prolonged storage or temperature variations. To mitigate potential risks, such as the dreaded "bottle bombs", consider the following actions:

- Allow the mead to mature for at least 12 months before bottling can enhance stability.

- Consume the mead while it's relatively young—this bypasses long-term storage concerns.

- Keep the mead in a consistently cool environment—this helps prevent unexpected fermentation.

- Should any bottle show signs of carbonation, promptly refrigerate the remaining bottles and consume them to avoid pressure build-up.

This approach to mead-making underscores a balance between art and science, where understanding the yeast's capabilities and carefully managing the fermentation environment can yield a mead of desired sweetness, without artificial interventions.

Implementing Pasteurization in Mead-Making for Stabilization

Pasteurization is a notable method for stabilizing mead, particularly favored for crafting sweet, carbonated variants. It's an alternative worth considering when other stabilization techniques are unavailable, or unsuitable for still mead. However, embarking on the pasteurization journey comes with its set of challenges. Inadequate pasteurization, the potential for boiling the mead, and the risk of over-pressurizing bottles require your utmost attention. It's noteworthy that boiling points can vary with alcohol content, occurring as low as 180°F in high-ABV meads and up to 190°F in those with lower ABV. A critical aspect of successful pasteurization is ensuring uniform sugar distribution and pressure within the bottles alongside closures capable of withstanding the induced pressure.

Prioritizing safety by utilizing bottles free of damage is essential, as even minor imperfections can lead to catastrophic failures.

- Begin by gently pre-heating the sealed bottles in a hot tap water bath, around 100–120°F, to minimize thermal shock and reduce the likelihood of breakage.

- For the pasteurization, heat a large pot of water to 160°F on the stove, ensuring enough water to submerge the bottles up to their necks. Alternatively, a sous vide device can simplify this process, allowing for a lower and more controlled temperature of 145°F. The precision and safety offered by sous vide technology mitigate many pasteurization challenges.

- Keep a close eye on the temperature of a "control" bottle. Achieving an internal mead temperature of 145°F for 20 minutes is sufficient to cease fermentation. This meticulous temperature control is crucial for halting microbial activity without compromising the mead's quality.

- Once pasteurized, carefully remove the bottles and let them cool. It's prudent to store them where any potential pasteurization failures won't pose a safety hazard. Conducting a carbonation check on a sample bottle after a few days is advisable to ensure the absence of "time bombs" in your batch.

Adopting safety measures such as wearing protective glasses, long sleeves, pants, and gloves is advised to safeguard against accidents. It's crucial to be vigilant about the environment you're working in, as the risk of bottles popping surely exists. Should you opt for the stovetop method, remember to remove the pot from the heat to prevent the residual heat from further increasing the water's temperature.

Back-Sweetening Your Mead

Sprinkling a bit of sweetness into your mead after fermentation is like adding the final touch to a masterpiece. Remember the golden rule: your mead needs stability. This step is crucial—it's the foundation that ensures your efforts in adding sweetness will genuinely enhance your brew, rather than lead to unexpected fermentations.

There is a right moment to introduce more sweetness. Picture your mead becoming clearer, almost like it's ready for its close-up. This clarity isn't just for looks; it ensures that the flavors you're about to enhance come unobstructed by fermentation remnants. The perfect time for this sweetening endeavor is soon after you've moved your mead away from the lees—preventing any unnecessary mixing of settled particles.

If the sweetness level you aim for feels a bit mysterious, here's a little trick: start with a generous sample from your batch and sweeten it gradually, observing changes with a hydrometer. This process is not unlike an artist mixing paints to get the perfect shade, except here, you're blending samples to hit that "sweet spot" of flavor (pardon the pun!). For example, by mixing samples with specific gravities of 1.000 and 1.030 in equal parts, you'll land on gravity of 1.015, a nice mid-point of sweetness.

Tasting these variations is where the magic happens. Trust your palate, but if you're stuck between choices, a simple palate can work wonders—or even better—a second opinion from a friend can provide that clarity.

When pinpointing your desired sweetness, some math will guide you to how much more honey you need to bring your vision to life. The formula...

$$\frac{Desired\ Gravity\ Increase}{0.035 * batch\ size\ in\ gallons} = X\ pounds\ of\ honey$$

... becomes your best friend. For instance, are you aiming to adjust a 5-gallon batch from a specific gravity of 1.000 to 1.025? You'll find yourself needing approximately 3.6 pounds of honey.

However, not all honey is created equal. Its natural moisture and sugar content variance means starting cautiously, with maybe half or three-quarters of your calculated honey addition, is a wise start. This way, you can inch closer to your desired sweetness without overshooting. After all, *adding* more sweetness is a breeze, but dialing it *back* is another story!

The Process

Blending honey into your mead requires a touch of finesse and a good deal of patience. It's a bit like a slow dance where gentle moves are appropriate. Avoid the temptation to stir too vigorously—creating a whirlpool or agitating too much liquid can introduce an unwelcome amount of oxygen into your brew. This can have unintended consequences on the final flavor profile you're working hard to achieve. For a smoother integration, consider employing a wine degasser, but remember, keep the speed low and steady.

A little trick to ease the honey into your mead more effortlessly is to warm it up first. Placing the honey container in a hot tap water bath softens it, making it less mixing-resistant. Thoroughly dry off the container afterward to prevent non-sterile water from accidentally mixing into your batch. Alternatively, mix the honey with a smaller quantity of mead in a separate vessel, giving it a more vigorous stir before combining it with the main batch. This technique can help ensure a uniform distribution without overly aggressive stirring in the primary fermenter.

Patience does pay off in this process. Don't worry if you notice a few stubborn honey globules that don't want to dissolve immediately. These holdouts will naturally dissolve into the mead over a few days, adding to the depth and complexity of its flavor.

After reaching your target final gravity, the next step is to allow your mead to clear. This period is crucial for conducting any final adjustments, such as fining or filtration, to refine the clarity and taste of your mead. It's also a good practice to monitor the specific gravity during this time to ensure that fermentation has definitively ceased, guaranteeing the stability and character of your mead remain intact.

Back-Sweetening Without Stabilization

This approach draws upon understanding the natural limits of your yeast and using them to achieve the perfect balance of sweetness. This method's cornerstone is introducing more honey than the yeast can ferment, effectively reaching the yeast's alcohol tolerance threshold. This technique allows the residual sugar to impart the desired sweetness to the mead.

- **Front Loading: A Delicate Balance:** Starting your fermentation journey with a heavy hand of honey sets the stage for what's to come. This process, known as front-loading, challenges the yeast right from the get-go, immersing them in an environment rich with sugar. While this method can push the yeast to produce a sweeter mead by reaching their alcohol tolerance quickly, it's not without its hurdles. The dense sugar environment can stress the yeast, potentially leading to unwanted flavors or the production of fusel alcohols. It's a balancing act, requiring a keen eye to ensure the yeast can thrive without overextending its capabilities. Surprisingly, yeast often defies expectations, fermenting beyond its limits, which might call for additional honey to hit the sweetness target.

- **Step Feeding: Precision and Patience:** Step feeding offers a more gradual route to sweetness. This method involves adding honey in stages, allowing the yeast to consume the sugar at a controlled pace until they reach their alcohol threshold. Then, additional honey is added to fine-tune the sweetness. This approach demands good attention to the yeast's behavior, which sometimes exceeds their anticipated alcohol tolerance, leading to meads with a higher ABV than initially planned.

Non-Fermentable Sugars

Introducing non-fermentable sugars into your mead offers a straightforward solution to achieving the desired sweetness without reactivating fermentation. This approach sidesteps the yeast's role in converting sugars to alcohol, allowing these sweeteners to blend seamlessly into your brew, enhancing its flavor without altering its alcoholic strength.

Among the popular choices are:

- **Erythritol and Xylitol:** Both sweeteners are excellent for adding that needed sweetness. However, a word of caution: xylitol is known to be harmful to dogs, so handle it with care.

- **Stevia and Splenda:** These are ideal for those looking to maintain the flavor profile without adding fermentable sugars, keeping the mead's character authentic.

- **Lactose:** Not only does lactose sweeten your mead, but it also enriches its body, lending a creamier texture and a fuller mouthfeel.

- **Maltodextrin:** Similar to lactose, maltodextrin can enhance the body of your mead, offering a more substantial sensory experience.

Utilizing non-fermentable sugars is a testament to the craft of mead-making, allowing for precision in crafting a brew perfectly tailored to your taste.

Perfecting Sweetness with Pasteurization of Bottled Mead

For those who prefer the traditional sweetness of honey but want to avoid re-fermentation, pasteurization is a technique worth exploring. As covered previously, the process involves gently heating the mead to a specific temperature for a certain duration, effectively stopping the yeast in their tracks without compromising the mead's flavor or alcohol content. The difference is that, in this case, the pasteurization process happens when the mead has already been bottled.

The process is delicate:

Bottling the mead, leaving the caps off, securing it with tin foil, and then carefully placing the bottles in boiling water is common. The critical temperatures for effective pasteurization are:

140°F (60°C) for 20 minutes

150°F (65°C) for 5 minutes

Adhering to these temperatures is paramount. Going beyond these limits risks evaporating the alcohol, diminishing the very *essence* of your mead.

Achieving Clarity

A mead that sparkles with clarity looks inviting and exemplifies the effort and care poured into its creation. On the other hand, a mead that's *less* than clear might raise eyebrows, leaving enthusiasts wary of sediment, questioning its craftsmanship. But clarity's role extends beyond first visual impressions—intricately linked to the mead's flavor profile and overall balance.

Time

They say "patience is a virtue", and in the world of mead-making, it's a virtue that rewards generously. Natural clarification is as old as mead-making itself, relying on the gentle passage of time to allow CO_2 and suspended particles to part ways. This isn't a quick fix but a testament to the mead-maker's dedication to quality. While the temptation to expedite this process with tools and techniques exists, such haste comes with its caveats—including the risks of oxidation or contamination—reminders of the delicate balance between art and science in mead-making.

The Challenge of Pectin Haze

Fruit meads bring their unique challenge to the quest for clarity: pectin haze. This stubborn veil obscures the mead's brilliance due to pectin not fully broken down during fermentation. Enter the hero of our story: pectic enzyme. This ally not only dispels the haze but also deepens the mead's color and enriches its flavor, extracting every nuance from the fruit used. While alcohol can dampen the enzyme's effectiveness, adding it at the fermentation's outset usually does the trick. For the diligent mead-maker, pre-fermentation maceration with the enzyme offers an even *greater* reward, ensuring every bit of potential haze is addressed before it can take hold.

The strategic addition of enzymes plays a crucial role in achieving that sought-after clarity. Since their power wanes in the presence of alcohol, timing their introduction is critical. Introducing enzymes alongside your yeast sets the stage for a clearer mead, minimizing the need for adjustments later. For those who find themselves adding enzymes post-fermentation, a double dose might be necessary, but it's a small price to pay for perfection!

Fining Agents

Achieving this desired clarity often involves using fining agents:

Bentonite

At its core, bentonite is a fining agent extraordinaire, adept at removing unwanted proteins from your mead through adsorption. It comes in two primary forms: sodium bentonite and calcium bentonite, each with unique characteristics and suitability for different mead-making scenarios.

- **Sodium Bentonite:** This variant is known for its assertive fining capabilities, making it ideal for a more vigorous action. It's particularly effective in higher pH environments, where its aggressive nature can shine without overwhelming the mead.

- **Calcium Bentonite:** A gentler counterpart, calcium bentonite is the go-to for lower pH conditions. It works swiftly to clarify your mead, forming compact lees that are easy to separate, thereby minimizing the risk of over-fining.

The choice between sodium and calcium bentonite hinges on the pH of your mead, with each type offering optimal performance under specific conditions.

Effective use of bentonite is all about the correct dosage and timing. Traditional methods advocate for its addition post-fermentation, but incorporating bentonite during fermentation presents a unique advantage. This approach allows the natural fermentation process to keep the bentonite in motion, enhancing its fining action while reducing the risks associated with post-fermentation handling, such as oxidation.

- **Dosage Matters:** Adhering to a dosage range of 2–6 grams per US gallon (0.5–1.3 grams per liter) strikes the perfect balance, ensuring clarity without compromising the mead's integrity. Overusing bentonite can lead to undesirable outcomes, perhaps a muted color palette or an unintended earthy note.

- **Rehydration for Optimal Results:** Rehydration is crucial if you add bentonite after the primary fermentation. Properly rehydrated bentonite is more effective, ensuring that it works as intended when you stir this slurry into your mead without leaving anything behind but clarity.

Not all bentonite is created equal, especially in mead-making. Opting for a product specifically designed for brewing ensures that you're using a food-grade agent that's safe and effective. The variety of bentonite available to home brewers is vast, but starting with a product from a reputable supplier can make all the difference in your mead's clarity and overall quality.

Gelatin

While finding gelatin on the grocery shelf is always possible, those serious about their mead know that the more refined, winemaking-specific gelatins bring superior results. These gelatins, primarily derived from animal sources, are formulated to flocculate with unmatched efficiency, leaving behind a mead that's visually appealing and free from haze.

For those venturing into commercial mead-making, vegetable-derived gelatins exist, though they are sold in bulk, making them less accessible for the hobbyist. This reality places animal-derived gelatin as the primary option for most, ensuring that clarity in your mead is within reach, regardless of scale.

Gelatin excels not only by clarifying mead but also by its ability to refine it—removing excess tannins and softening the color. This capability is particularly useful in young meads, where the boldness of tannins and color might need tempering. However, this power necessitates careful dosing—too much can strip away the essence that gives your mead its character.

For standard clarification, a dosing guide suggests 75 to 550mg per gallon (20–150 mg/L). If you intend to adjust tannins and color, pushing the dose up to 2000 mg per gallon (530 mg/L) is an option, albeit one that requires precision to avoid overcorrection.

Introducing a counter-fining agent like bentonite or Kieselsol shortly after the gelatin application is a smart move to ensure that gelatin doesn't overstep its role. This step acts as a safeguard, deactivating the gelatin to prevent over-fining and helping to clear out any remaining positively-charged particles.

The effectiveness of gelatin depends on its preparation. Dissolving it in hot water—at the right temperature and volume—ensures it integrates smoothly into your mead. Stirring the mead as you add the gelatin guarantees an even distribution, setting the stage for a beautifully clear final product.

Isinglass

Isinglass, a fining agent steeped in tradition and efficacy, offers mead-makers a nuanced approach to refining their brew's clarity and mouthfeel. Derived from the swim bladders of fish, this collagen-based agent is an example of the innovative ways natural resources can be harnessed in the craft of mead-making.

Isinglass caters to a variety of preferences and methods, presented in three primary forms:

- **Prehydrolized Isinglass:** This powder form, known for its high refinement, requires rehydration in cold water for approximately 30 minutes before use. Its precision in clarifying mead makes it a favored choice among mead-makers aiming for perfection in their craft.

- **Flocced Isinglass:** In its sheet form, flocced isinglass demands a more extended rehydration period of at least 24 hours in cold water, occasionally needing an adjustment in the water's acidity to achieve the best results.

- **Liquid Isinglass:** Offering the utmost convenience, liquid isinglass eliminates the rehydration step, allowing direct addition to the mead for those seeking efficiency without compromising quality.

The use of isinglass extends beyond mere clarification. It is celebrated for enhancing the mead's mouthfeel, smoothing any rough edges that may detract from the drinking experience. By targeting the polyphenolic compounds responsible for astringency, isinglass softens the mead without the aggressive impact seen in other fining agents like casein or gelatin. This gentle approach also aids in unveiling the subtle fruit notes that might remain hidden.

Isinglass shines in nearly clear meads, producing fluffy lees prone to clinging to the sides of the aging vessel. A meticulous racking process, complemented by counter-fining with agents like Kieselsol, can be invaluable in mitigating this. This not only compacts the lees but also ensures that the fining process does not overstay its welcome.

For meads with lower tannin levels, a preparatory addition of powdered tannins before the isinglass can optimize its effectiveness, leveraging its need for tannins to work its magic.

When incorporating isinglass into your mead, staying alert to any off-odors, especially a fishy smell, is crucial, as this indicates spoilage. Furthermore, adherence to the recommended dosing, which varies by form, ensures that the isinglass performs its role without diminishing the mead's inherent qualities.

Chitosan

Originating from chitin, often sourced from the outer shells of crustaceans, chitosan's positive charge makes it good at drawing out negatively-charged particles, thus clarifying the mead without altering its essential qualities.

For those of us dedicated to home-brewing, (me and you!) it's reassuring that our chitosan is carefully refined, distancing itself from the proteins that trigger shellfish allergies. Interestingly, in the U.S., commercial winemaking adheres to stringent guidelines, necessitating that chitosan comes from the mold *Aspergillus Niger*—a move ensuring allergen-free clarity enhancement. With its recent approval as a clarifying agent, *Aspergillus*-derived chitosan is on the cusp of broadening our fining toolset, promising a new era of clarity in mead-making.

Beyond its clarifying prowess, chitosan is celebrated for its antimicrobial properties. This added benefit is invaluable in winemaking, aiding in the battle against spoilage culprits like Brettanomyces and lactic acid bacteria. For mead-makers, this means achieving a visually stunning brew and enhancing its stability and longevity.

Chitosan's independence from tannins sets it apart, making it a perfect match for mead, which often lacks these compounds. This flexibility, coupled with its ability to form compact lees, simplifies the racking process, steering clear of the challenges of other fining agents that leave behind fluffy, stubborn lees.

Incorporating chitosan into your mead necessitates a gentle touch and an adherence to recommended dosages, typically ranging from 5.5–7.5 ml per gallon for a 1% solution. This careful approach guarantees that the mead's integrity remains intact, highlighting the artisan's commitment to quality at every step.

Kieselsol (Silica gel)

Kieselsol, a fining agent derived from silica gel or silicon dioxide, stands out for its exceptional ability to refine mead.

Kieselsol operates on a remarkably straightforward principle: it carries a strong negative charge that attracts and binds positively-charged particles, including those pesky proteins that cloud the mead. This process enhances the clarity of the mead and preserves the integrity of its taste and aroma, ensuring that what you *love* about your mead remains untouched. The result?

Very compact lees facilitate a cleaner, more efficient separation process, minimizing waste and preserving more of your precious brew.

Beyond its standalone prowess, kieselsol shines as a counter-fining agent. Its talent for compacting the fluffy lees left by others makes it an indispensable ally in achieving a polished, sediment-free finish. This characteristic is invaluable, especially when working with fining agents that, while effective in their primary role, leave behind a less desirable residue.

The magic of kieselsol lies not just in its properties but in how it's used. With a typical concentration of 30%, the dosing sweet spot is between 1–2 ml per gallon, a range that ensures optimal performance—without the risk of overdoing it. Adherence to this dosing guideline is critical, providing the mead's clarity is enhanced without altering its core characteristics.

Combining Kieselsol and Chitosan (SuperKleer/DualFine)

The pairing of chitosan and kieselsol, often found together in products like DualFine or SuperKleer, is a powerful solution for those pursuing this clarity. This duo has garnered acclaim for its swift and effective results, transforming even the murkiest meads into shining beacons of transparency.

The strength of chitosan and kieselsol lies in their complementary actions. Kieselsol, carrying a strong negative charge, initiates the process by latching onto positively-charged particles that cloud the mead. Conversely, chitosan brings a positive charge to the mix, targeting the remaining negatively-charged particles. Together, they form a net that captures a broad spectrum of impurities, leaving behind a clear and vibrant mead.

While the efficacy of chitosan and kieselsol is undeniable, practical considerations can shape the mead-maker's experience. The standard packaging of DualFine pouches is tailored for 5-6 gallon batches, posing a challenge for smaller-scale endeavors. The inconvenience of adjusting these pouches for lesser volumes and the cost implications invite a closer look at alternatives.

For those crafting larger quantities of mead or seeking a more economical approach, buying chitosan and kieselsol separately by the liter is as a savvy choice. This route offers a cost advantage—mirroring the expense of about eight DualFine pouches while treating up to 150 gallons—and grants mead-makers greater control over dosing. Though the recommended dosing for the combined use of these agents is on the higher side, tailoring the dosage to the specific needs of your batch can lead to flawless clarity, without overfining.

The Magic of Egg Whites in Mead

Egg whites, thanks to the albumin they contain, act as a natural, positively-charged fining agent. Their real power shines in tackling harsh tannins, making them a prime choice for meads with the boldness and depth of heavily fruited varieties. This approach, borrowed from the age-old winemaking traditions, suggests that egg whites that parallel the tannic intensity found in red wines are most beneficial for meads.

For optimal use, combine an egg white with a pinch of salt and a few drops of water, mixing gently, to avoid froth formation. This preparation should then be dosed into your mead at 0.5–1ml per gallon. Following this addition, a waiting period of 1–2 weeks allows the egg white to fully interact with the mead, after which you can rack off the resulting sediment.

Casein: A Protector Against Oxidation

Casein, predominantly found in milk, is a guardian against the phenolic compounds responsible for oxidative damage, which can impart bitterness and browning to your mead. While it's not the total solution for oxidation, casein softens its impact, subtly enhancing the mead's sensory profile. Its clarifying impact may be modest, but its value in improving mead quality is undeniable.

In practical applications, casein is often introduced in forms like potassium caseinate or as part of proprietary blends such as Polycacel, which combines PVPP, casein, and cellulose for more robust fining action. For those seeking a straightforward method, skim milk, rich in casein, can be added to the mead at a dosage of 55–114 ml per gallon (15–30 ml/L); ensure continuous stirring for even distribution.

PVPP

Polyvinylpolypyrrolidone (PVPP) is a synthetic fining agent that brings precision to this process. Derived from Nylon 6, PVPP's role in mead-making is as nuanced as it is practical, offering a refined approach to tackling oxidation challenges, without compromising the mead's sensory qualities.

What sets PVPP apart is its method of action. This agent is insoluble in mead, allowing it to specifically target and adsorb low-weight phenolic compounds through targeted chemical bonding. This action is crucial for combating oxidative damage—phenolics that can lead to browning, bitterness, and other off-flavors detracting from the mead's quality. Remarkably, while PVPP excels in clarifying the mead, it does so without altering the aroma.

PVPP serves a dual purpose in the mead-making process. As a prophylactic agent, it preemptively binds to the precursors of oxidative compounds, effectively acting as a shield against future damage. PVPP addresses existing oxidative issues when treated, restoring the mead's intended clarity and taste profile. This versatility makes PVPP an invaluable tool in the mead-maker's arsenal for maintaining the integrity of their creations over time.

In the U.S., commercial mead products incorporating PVPP are subject to TTB regulations requiring filtration to eliminate any residual particles. While home brewers are exempt from this mandate, increasing awareness about microplastics has prompted some to consider the implications of introducing PVPP into their mead. However, the design and application of PVPP—to precipitate out of suspension—combined with diligent racking practices significantly reduce the likelihood of any residual particles remaining in the final product.

The key to harnessing PVPP's benefits lies in accurate dosing, which varies according to the product's grade. Adherence to the manufacturer's guidelines is essential for achieving the desired clarity without overfining. Brands like Polyclar VT suggest dosing ranges from 0.5 to 2.5 grams per gallon, tailored to preventative or corrective measures against oxidation.

Sparkalloid

Sparkalloid carved out its niche in the world of mead-making as a distinguished fining agent, thanks to its unique composition and the clarity it brought to countless batches. Crafted by Scott Labs, this blend of polysaccharides and diatomaceous earth became a favored choice among artisans for its gentle, yet effective clarifying power, leaving the mead's essential characteristics—color, flavor, and aroma—untouched.

Sparkalloid's presence changed significantly between 2020 and 2021 when Scott Labs discontinued its production. The health risks associated with inhaling diatomaceous earth, both in its creation and application, prompted this decision. Despite this shift, Sparkalloid's presence lingered in the home-brewing community, a testament to its efficacy and the trust it garnered over time.

What set Sparkalloid apart was its robust positive charge, making it a powerhouse in removing particulates that clouded the mead, resulting in a brilliantly clear finish. Its reputation for quick settling, and producing fluffy, easily disturbed lees highlighted the need for a thoughtful approach to racking. Using Sparkalloid with agents like bentonite or kieselsol was often recommended to achieve more compact lees, streamlining the clarification process.

Sparkalloid offered two formulations—cold mix for juices, and hot mix for wines and meads—catering to a range of clarification needs. The dosing guidelines, suggesting a range of 150 to 500 PPM (approximately 0.6–1.9 grams per gallon), allowed for flexibility in application. Preparing Sparkalloid involved creating a 2% slurry and boiling it for at least 15 minutes to fully activate its fining properties. This careful preparation, followed by thorough stirring into the mead, ensured Sparkalloid could work its magic effectively.

Gum Arabic

Gum Arabic, a fining agent with a rich heritage, is derived from the sap of acacia trees, native to Africa. Its role in mead-making is multifaceted, acting as an adsorbing protective colloid that gracefully preserves the integrity of colloidal matter. This unique property positions Gum Arabic as an indispensable fining agent, best applied in the final stages before bottling to ensure it complements, rather than competes, with other fining processes.

What sets Gum Arabic apart is its capability to safeguard the colloidal state of mead, preventing the flocculation and subsequent sedimentation of particulate matter. This is particularly valuable for young red wines or meads, where it stabilizes anthocyanins—those delicate pigments that gift color, but are prone to instability.

Gum Arabic's influence extends well beyond color stabilization. It's a harmonizer, adept at softening the bite of astringent tannins and smoothing the sharpness of acidity, all while adding to the body and mouthfeel of the mead. This enhancement of sensory attributes makes Gum Arabic a favored choice for those seeking to round out their mead's profile for immediate enjoyment.

For meads destined for a sparkle, Gum Arabic plays yet *another* role. Its addition can decrease surface tension, amplifying the fizziness in carbonated meads and elevating the effervescence to delight with every sip.

The method is as important as the measure when incorporating Gum Arabic into mead. Typically available as a 20% solution for winemaking, the dosing ranges from 2 to 7.5 ml per gallon (0.5 ml/L to 2 ml/L). It's crucial to adhere to the manufacturer's directions, as concentration levels can vary. While winemaking-specific Gum Arabic is designed to be filter-friendly, caution is still advisable with alternative preparations to avoid potential clogging issues.

Carbon

With its robust fining capabilities, activated carbon is an excellent ally in refining mead, targeting unwanted colors, aromas, and flavors. Its application, however, demands a discerning touch. Given its non-selective nature, activated carbon can inadvertently remove *desirable* characteristics alongside undesirable ones. This fact underscores the importance of conducting careful fining trials before committing to the full application in your mead production.

When choosing activated carbon, knowing the nuances between the available types is critical. There are primarily two forms: one that excels in removing color, often denoted as KBB, and another that is more effective in eliminating unwanted aromas, usually marked as AAA. This distinction is vital because it allows marketers to target specific issues precisely. Opting for activated carbon designed explicitly for wine-making (and mead-making) is necessary, as generic variants can lead to unpredictable outcomes—making the fining process more of a guessing game than a science!

Determining the correct amount of activated carbon to use is an exercise in balance. The dosing will vary based on how the carbon has been processed, so following the manufacturer's recommendations is non-negotiable for achieving optimal results. Too little, and you may not address the issue at hand; too much, and you risk stripping away the very essence that makes your mead unique. It's common to pair activated carbon with another fining agent, such as PVPP, to mitigate its broad effects. This counter-fining approach helps refine the mead's profile, without sacrificing its character.

Cold-Crashing

Cold-crashing is often discussed in home-brewing circles, offering a method to en-hance the clarity of mead by leveraging low temperatures. This technique, however, comes with its caveats and is not a one-size-fits-all solution.

At its core, cold-crashing involves chilling the mead to refrigerator temperatures, en-couraging the flocculation of specific yeast strains thus aiding in particle settlement. It's essential to say that cold-crashing is not a means to stabilize or eliminate yeasts, but rather a method to expedite the clarification process by keeping CO_2 dissolved, and facilitating the descent of particles out of suspension.

The effectiveness of cold-crashing largely depends on the type of yeast used in the mead. High-flocculating yeasts, typically found in ale brewing, respond well to cold temperatures, making them more likely to flocculate and settle. In contrast, many yeasts used specifically for mead-making are of the low-flocculating variety and may not show significant improvement with cold-crashing alone.

Additionally, mead-makers should be wary of the potential risks associated with this technique, particularly the increased likelihood of oxygen exposure and airlock suck-back. The latter occurs as the fermenter's internal pressure decreases with the temperature, posing a risk of drawing airlock contents back into the mead, which could introduce oxygen or contaminants.

While cold-crashing can serve as a quick clarifying aid, it's time that often plays a pivotal role in clearing the mead and enhancing its flavor.

Extended cold conditioning, much like the lagering process in beer brewing, emerges as a more effective strategy for achieving a crystal-clear mead. This method involves keeping the mead at cold temperatures for several weeks to months, allowing even the finest particles to settle gradually without the haste and potential risks of cold-crash-ing.

Filtration

Filtration is a refined technique in mead clarification, demanding a blend of specialized equipment and meticulous process. It's not an indispensable step—given the efficacy of readily available fining agents—but filtration emerges as a premier choice for those aiming for unparalleled clarity in their mead.

Success in filtration depends on the choice of equipment. Everyday items like coffee filters and cheesecloths prove inadequate, failing to effectively reduce lees and posing a significant risk of mead oxidation.

Specialized filtration units, designed with precise pore sizes measured in microns, offer a targeted solution by specifying the maximum particle size. This level of specificity is pivotal for achieving a clear and polished mead.

Filtration technology categorizes filters into two main types: absolute and nominal. Absolute filters promise a comprehensive removal of all particles exceeding their micron rating, delivering on the promise of clarity. Nominal filters, while effective, allow for a slight margin of error, with some particles potentially eluding capture. This distinction is crucial for mead-makers to consider, ensuring expectations align with the filter's capabilities.

The Buon Vino Minijet is a favored option for homebrew enthusiasts exploring filtration. This plate-and-frame style filter system supports three grades of filters, adaptable to the mead's initial clarity. A #1 filter is recommended for tackling cloudiness, followed by a progression to higher grades for additional refinement. Should the mead already present as mostly clear, initiating the process with a #2 filter often suffices. However, it's essential to recognize that the #3 pads, despite being marketed as "sterile" act as nominal rather than absolute filters. Therefore, they do not ensure mead stabilization or safeguard against refermentation after back-sweetening.

Other Considerations: Counter-Fining

The technique of counter-fining is a sophisticated strategy which employs a combination of fining agents to target and remove a broad spectrum of sediments. This approach, coupled with a mindful consideration of the quantities used to avoid over-fining, will mean the mead's character remains intact.

Counter-fining is a systematic approach involving using two or more fining agents, typically with opposing charges, to address the diverse particles in mead. This ensures a more effective clarification process. Key strategies include:

- **Combining Bentonite with Sparkalloid, Isinglass, or Chitosan:** Initially utilizing Bentonite during fermentation helps clear positively-charged particles. Subsequently, introducing a negatively-charged agent such as Sparkalloid, Isinglass, or chitosan after racking enhances the removal of lingering sediments, capitalizing on the interplay of opposite charges for a thorough clean.

- **The Synergy of Kieselsol and Chitosan:** This duo, often presented as DualFine/SuperKleer, is a testament to the efficacy of counter-fining. Kieselsol's ability to attract positively-charged particles is complemented by chitosan's affinity for negatively-charged elements. Kieselsol bolsters chitosan's effectiveness when used in succession, accelerating sediment removal and achieving a swift clarification.

The Importance of Moderation in Fining

The work to achieve clarity is fraught with the risk of over-fining, where the zealous removal of sediments can strip away the mead's vibrant color and rich flavors. To navigate this:

- **Exercise Restraint:** Begin with the lowest recommended dosage of fining agents, adjusting only as necessary. Conducting small-scale bench trials offers a safe way to determine the optimal amount needed for your specific batch, reducing the risk of compromising the mead's quality.

- **Approach Carbon/Charcoal with Caution:** Recognized for its potent fining capability, carbon or charcoal should be considered a measure of last resort due to its propensity for over-fining. Exploring alternative fining agents often yields satisfactory results without the significant risk of diminishing the mead's natural attributes.

Kegging

This method, favored for its straight-forwardness, offers various benefits to those mead-makers looking to streamline their process, particularly when carbonating their creations.

The simplicity it brings to carbonation and packaging is at the heart of kegging's appeal. By transferring mead directly into a keg, sealing, and attaching it to a CO_2 tank, mead-makers can bypass the labor-intensive steps associated with bottling—no more meticulous sanitizing of bottles, no capping, and no corking!

Serving mead becomes more convenient, allowing individual glasses to be poured without the pressure to consume the entire bottle promptly, minimizing oxidation concerns. For gatherings, kegging eliminates the hassle of multiple open bottles, offering a single, continuous mead source. Moreover, the sustainability factor of kegs, which can be used indefinitely, adds an eco-friendly aspect to this method.

Ball Lock Keg/Credit: Svarun (www.Shutt erstock.com)

However, kegging is not without its initial hurdles. The upfront cost of purchasing kegs can be a significant investment, deterring those on a tight budget. The system's dependency on CO_2 also means that a depleted CO_2 supply can halt mead dispensing until a refill is obtained. Furthermore, sharing mead in a social setting loses the tactile and ritual charm of passing around bottles. For enthusiasts of forced carbonation, the necessity for additional chilling equipment like a kegerator or keezer adds another layer of expense and consideration, especially given the critical role temperature plays in achieving the desired carbonation levels.

Beyond serving, kegs offer versatility in bulk aging, allowing the mead to be aged under a blanket of CO_2 or nitrogen to stave off oxygen and potential oxidation. This method supports dispensing mead in controlled amounts over time—without compromising the batch's integrity.

However, kegging does require a careful understanding of the equipment and process involved. You need more than a cursory understanding; it's a comprehensive aspect of mead-making that encompasses considerations of cost, equipment, and the nuances of carbonation and preservation. For those intrigued by the prospects of kegging, engaging in thorough research and seeking advice from seasoned mead-makers or local homebrew supply stores can illuminate the way forward.

Given the intricate nature of kegging and the wealth of specific knowledge required to do it correctly, it's clear that a summary would not do justice to the subject. With its detailed processes and technical requirements, kegging is a topic that genuinely benefits from in-depth exploration and discussion. Therefore, I recommend thoroughly researching and engaging with the rich resources available on homebrewing websites, forums, and blogs. These platforms offer comprehensive guides, step-by-step tutorials, and lively discussions that can provide a thorough understanding of kegging from setup to maintenance.

Additionally, visiting your local homebrew shop can be immensely beneficial. The expertise of the staff and the opportunity to connect with other homebrew enthusiasts can offer invaluable insights and even hands-on demonstrations tailored to your specific needs and queries. This combination of self-directed learning and community interaction will equip you with a solid foundation in the theory and practical know-how to confidently begin your kegging journey!

Aging

The process of aging mead is rich with tradition and innovation, blending ancient practices with modern fermentation techniques. Contrary to the old belief that mead must age for years to reach its prime, contemporary methods allow for much earlier enjoyment. Yet, maturation remains a revered path to unlocking the fullest expression of mead's flavor.

Aging mead is like fine-tuning an instrument, where subtle adjustments can harmonize the composition. This maturation process softens the edges of any bitterness and alcohol sharpness, and gradually clarifies the mead by precipitating proteins, tannins, and other particulates.

The most striking benefit of aging is the seamless integration of flavors, creating a smooth, cohesive profile. While vibrant, youthful meads capture the heart with their boldness, their depth and complexity are often revealed through careful aging, especially in varieties enriched with tannins that further benefit from extended maturation.

Distinguishing Aging from Oxidation

Understanding that aging mead doesn't mean exposing it to oxidation is crucial. Proper aging is a controlled process aiming to enhance the mead without the adverse effects of oxygen exposure. Innovative techniques, such as wax-dipping bottles or selecting synthetic corks, protect against oxidation. Despite these precautions, a degree of oxidation is inevitable—yet when managed skillfully, it can add intriguing layers of complexity to the mead, rather than detract from its quality.

External Influences on Mead Aging

The aging process of mead is susceptible to external factors like heat, light, and physical disturbance, each accelerating chemical reactions affecting mead's character. Awareness and control of these conditions are paramount, as they can either hasten the unwanted aspects of aging or, in a controlled environment, contribute positively to the mead's development. For instance, when done precisely, intentional oxidation can introduce nice nutty and sherry-like notes reminiscent of traditional mead styles.

Deciding Between Bulk and Bottle Aging

The choice between bulk aging and bottle aging offers mead-makers flexibility based on their individual preferences and the specific qualities of their mead. Bulk aging allows for uniform maturation and easier adjustments before the final bottling, providing consistent quality across the batch. Bottle aging, conversely, invites the mead to evolve within its vessel, potentially developing unique characteristics influenced by the conditions of its environment. Each method, covered below, has its merits, with the decision reflecting the mead-maker's aspirations for their beverage creation.

Bulk Aging

Bulk aging is where the brew can mature in a collective vessel, typically a carboy, before being bottled.

Initiating the bulk aging process involves leaving the mead in a secondary container post-primary fermentation. Unlike immediate bottling, which can leave sediments in individual bottles, bulk aging ensures that yeast and other particulates have adequate time to flocculate—gather and sink. This natural clarification reduces the need for frequent racking, thereby lessening the mead's exposure to oxygen, a crucial aspect of preserving its quality and taste.

One of the foremost challenges in bulk aging is managing oxidation. To mitigate this, several strategies are employed:

- **Minimizing Headspace:** Selecting an aging vessel with minimal headspaces, such as a narrow-neck carboy, is vital. This reduces the surface area where the mead comes into contact with air.

- **Volume Planning:** Anticipating the volume of mead post-fermentation and planning the batch size can help ensure that the secondary container is filled appropriately, leaving little to no excess air.

When faced with unavoidable headspace, these creative solutions come into play:

- **Inert Gas Purging:** Filling the remaining space with an inert gas like CO_2, nitrogen, or argon can shield the mead from oxygen. This is accessible to those with kegging setups or can be achieved using food-grade CO_2 cartridges available online.

- **Topping Up:** Adding extra mead from a reserved portion of the batch or a similar mead helps eliminate headspace without diluting the brew's character.

- **Using Filler Materials:** as previously discussed, filling the excess space with sanitized, inert materials like marbles is a last resort option.

Bulk aging mead opens a window to deepen its flavors and enhance its quality. Yet, navigating through the aging process demands careful attention to oxidation—a formidable foe in preserving the essence of mead.

Below, we look at practical methods to safeguard your mead against oxidation, ensuring its aging through time *enriches* its character rather than diminishes it.

Embracing Sulfites for Oxidation Control

Introducing sulfites, such as potassium metabisulfite or Campden tablets, into your mead acts as a first line of defense against oxidation. These compounds release sulfur dioxide (SO_2), proactively seeking out and neutralising free oxygen, thereby protecting the mead's delicate flavors and aromas. Commonly embraced in the wine and commercial mead industries for their efficacy and safety, sulfites, when used in moderation (25–50 ppm is the sweet spot for most mead varieties), do not impart any off-flavors. Incorporating sulfites at each racking enhances your mead's defense against oxidation.

The Critical Role of Proper Equipment

The ways of transferring mead during bulk aging significantly impacts its exposure to air. A direct pour from one vessel to another can aggressively introduce oxygen, quickening the pace of oxidation. Opting for a siphon method, on the other hand, significantly *reduces* air contact. This technique involves filling the receiving container from the bottom up, minimizing the mead's interaction with air. Mastery of the siphon calls for a delicate balance, starting high and gradually lowering to avoid disturbing the sediment. Engaging two people for this task—one to manage each end of the siphon—can reduce potential mishaps and ensure a smooth transfer.

A Word of Caution on Filtration

While filtration can contribute to the clarity and stability of mead, using basic household items like coffee filters is ineffective and can exacerbate oxidation. True clarification requires specialized filtration equipment designed to handle mead without introducing oxygen. Thus, mead enthusiasts are discouraged from improvising with inadequate tools, which can compromise the beautiful mead's quality!

Determining the Duration of Bulk Aging

The appropriate length of bulk aging is influenced by the style of the mead and the desired outcome. While there's no one-size-fits-all timeline, aging durations can range from a few months to several years. The key is to align the aging period with the specific characteristics and flavor profile you aim to achieve in your mead.

Bottle Aging

Bottle aging is another significant phase in the life of mead, where it continues to evolve and mature—even after bottling. The aging process is influenced by several factors, including the type of closure used, with oxygen exchange playing a key role in the speed of aging. Here we explore the nuances of bottle aging, focusing on the impact of corks and how they can affect the longevity and quality of your mead.

Even in the absence of oxygen, as in the case of a crown-capped bottle sealed to near perfection, chemical reactions within the mead persist. These reactions contribute to the maturation of the mead, enhancing its complexity and flavor over time. However, when natural corks are used, the slight oxygen exchange can accelerate this aging process; choosing a bottle and cork significantly affects the outcome.

The Significance of Cork Selection

Choosing the right cork is more than a matter of fit; it's about balancing ease of use with the longevity of the seal. In the U.S., standard wine bottles are compatible with size #8 and #9 corks. While #8 corks might be easier to insert with a double-lever corker, they don't seal as tightly as #9, potentially shortening the mead's lifespan before oxidation sets in. The quality of the cork also plays a crucial role. Agglomerated corks, for example, may degrade within a few years, risking oxidation of the mead. Investing in higher-quality corks can significantly extend the life of your bottled mead, making those few extra cents per cork well worth it in the long run.

Techniques for Corking

Achieving an optimal seal demands precision in the corking process. Solid #9 corks offer the best seal but can be challenging to insert manually. A floor corker, while more of an investment, provides ease and consistency in corking. Some homebrew shops even offer rentals of such equipment. Additionally, minimizing headspace in corked bottles is essential for long-term aging. Perfecting this technique can dramatically reduce the amount of mead lost to evaporation through the cork over time.

Corks are not just another component; they're a critical element that requires attention and care. Corks should ideally be used within six months of purchase to ensure they retain their sealing properties. Proper storage and timely use of corks make a significant difference to the success of bottle aging.

Oaking

French Oak Wood Chips/Credit: boommaval (www.S hutterstock.com)

Oaking is a method embraced across all styles of mead, enriching the drink with complexity and a palette of flavors that speak to both the connoisseur and the curious. While introducing oak into mead can be at the maker's discretion, it's important to remember that the essence of oaking is subtlety and balance. Over-oaking can overshadow the mead's intrinsic qualities, like winemaking, where harmony between elements is desired.

Oak aging is not merely about imparting new flavors but elevating the mead to a realm of structured complexity and refined sensory experiences. Oak contributes over 70 volatile aroma and flavor compounds, weaving notes of vanilla, spice, and woodiness into the mead's character—each layer adding to the interesting narrative of the drink.

The appeal of oaking lies in its ability to introduce a structured backdrop against which the mead's flavors can unfold. The process:

- Infuses the mead with a vanilla sweetness and spiciness, creating an inviting warmth.

- Adds a subtle sweetness that enriches the mead's body, enhancing its savoriness without tipping the balance.

- Imparts a woody depth, grounding the mead with an earthy foundation.

- For pyments, especially those crafted with red grape varietals, a bolder oak character can align the mead more closely with traditional wine profiles, offering a bridge between the worlds of mead and wine.

Oak's contribution to mead extends beyond simple flavoring; it introduces a structural depth and rich sensory nuances, making your mead experience more engaging—with over 70 volatile aromas and flavor compounds attributed to oak.

Unveiling Oak's Flavor Palette:

- **Lactones Unveiled:** The whispers of cis- and trans-oak lactones emerge from the untoasted heart of the oak, painting the mead with strokes of coconut, floral, and earthy notes. This bouquet, reminiscent of the aromatic freshness of an oak grove, subtly enriches the mead's profile.

- **Complex Sugars and Aromas:** Through the natural aging and toasting of oak staves, furfural, and 5-methyl furfural unfold, transforming into simple sugars that lend body and evoke butterscotch and mocha aromas. These elements, accentuated by the smoky and toasty nuances at higher temperatures, add a warm, inviting depth to the mead.

- **Vanillins:** The transformative power of heat on lignin releases vanillins, which imbue the mead with a gentle vanilla flavor.

- **Spice Notes:** The interplay of eugenol and isoeugenol, derived from the essence of oak and its transformation under heat, introduces spicy, clove-like undertones. Once melded into the mead, these elements create a rich tapestry of flavors.

- **Smoky Depth:** The deeper toasting of oak beckons guaiacol and 4-methyl guaiacol into the fold, imparting a robust smoky character that deepens the mead's complexity, reminiscent of a fireside gathering under a starlit sky.

- **Sweetness and Structure:** While cellulose provides the oak with its sturdy frame, hemicellulose breaks down to release sugars, enriching the mead with caramelized sweetness and toasted flavors, enhancing its character without compromising its integrity.

- **Tannins:** The subtle strength of tannins, varying between American and French oak, introduces a foundational structure and plays a pivotal role in the mead's aging journey. Their sensitive nature to processing and environment influences the mead's maturation, adding a layer of sophistication to its development.

Selecting the Right Oak for Your Mead

A myriad of oak species offers mead-makers a rich palette. Let's explore the key considerations in choosing the right oak for your mead.

- **American White Oak** (*Quercus alba*) stands out for its quick flavor infusion, imparting a distinctive blend of vanilla, woody, and toasty notes. Its appeal among mead-makers is bolstered by its affordability, broad availability, and diverse toast levels, making it a versatile choice for various mead styles.

- **European Oaks**, including Q. robur, Q. Petraea, and particularly Q. sessilis, are esteemed for their contribution to the depth and elegance of oak-aged beverages. Each species adds a layer of sophistication and complexity, making them a prized choice for mead-makers aiming for a refined oak character.

Given barrel usage's practical and economic constraints, home mead-makers typically opt for oak as chips, cubes, dominoes, and staves. This section covers the three most accessible types of oak:

- **American Oak** is favored for its pronounced vanilla and woody character, quickly enriching mead with its bold profile. Its cost-effectiveness and wide range of toast levels make it a go-to option for adding a rich, sugary, and toasty dimension to mead.

- **Hungarian Oak** offers a more subtle and gradual flavor infusion, characterized by its fine grain resulting from slower growth. This oak variety gently imparts toasted, vanilla, and caramel-like flavors, providing a sophisticated touch suitable for meads that benefit from a delicate oak presence.

- **French Oak** represents the pinnacle of oak aging and is known for its high tannin content and rich flavors, including sweet, woody, and spicy notes. While it commands a higher price, the complexity and depth it adds to mead are unmatched, ideal for crafting meads that can stand up to its intense character.

Understanding Toast Levels

The toasting level of oak significantly influences the flavors and aromas imparted to mead:

- **Light Toast** introduces a gentle coconut or vanilla character, ideal for dry meads seeking a subtle oak nuance.

- **Medium Toast** enhances the mead's aroma more than its flavor, offering a warm, sweet essence with pronounced vanilla notes well-suited for traditional meads and light models.

- **Medium Plus Toast** brings in flavors of honey, roasted nuts, and a hint of coffee, aligning perfectly with robust red pyments, bold melomels, and braggots.

- **Heavy Toast** delivers intense caramelized and charred flavors swiftly and should be used judiciously to avoid overwhelming the mead.

Incorporating Oak into Your Mead

Oak is available in many forms, each with unique characteristics and benefits. Chips and cubes are renowned for their accessibility and the broad spectrum of flavors they impart, making them a popular choice for mead-makers looking to experiment with oak's influence. Spirals and strips, offering a more controlled and gradual flavor infusion, are excellent for those seeking to enhance their mead with subtle complexity.

While oak powder and essence might seem convenient, however their intense and sometimes harsh flavors can detract from the mead's natural qualities and thus are generally not recommended. For those drawn to the traditional method, barrels provide a time-honored way to infuse mead with a deeply integrated oak character—albeit at a higher cost and requiring more space.

The interplay between the oak's surface area in contact with the mead and the duration of exposure is crucial in shaping the mead's oak character. Small-format oak products like chips offer a swift saturation, delivering a quick burst of flavor that, while effective, might lack the layered complexity many mead-makers desire. In contrast, larger formats such as cubes and spirals allow for a slower release of flavors, imbuing the mead with a richer, more intricate profile.

Adding oak post-fermentation is widely practiced, taking advantage of alcohol's natural antimicrobial properties. Some advocates of oak infusion prefer to sanitize their oak through steaming or boiling, though this might leech valuable flavors and tannins from the wood. A gentler approach involves rinsing the oak in a sulfite solution, ensuring cleanliness without sacrificing the wood's natural contribution to the mead. It's important to note that introducing oak into mead can result in additional particulates, which may necessitate further clarification to achieve visually appealing clarity.

Once oak has been added to the mead, its presence is permanent, potentially mellowing but not disappearing over time. This permanence underscores the importance of a cautious initial approach—beginning with minimal oak addition and adjusting incrementally allows for greater control over the mead's flavor development, ensuring a balanced and harmonious integration of oak.

Influence of Oak on Mead

The specific effects of oak on mead depend significantly on the type of oak chosen, its level of toasting, and the duration for which the mead is exposed to it. Let's explore how oak acts as a transformative element in mead, uniquely enriching its taste, color, and texture.

Elevating Taste and Aroma

Oak's contribution to mead is immediately noticeable in its rich flavors and aromas. Standard oak characteristics include distinctive oaky and woody notes, complemented by the comforting essence of vanilla and the robust depth of toastiness. When oak is more intensely toasted, it bestows even deeper flavors upon the mead, such as caramelized and charred nuances, adding complexity to the drink.

Enhancing Color Depth

Beyond taste and aroma, oak also plays a pivotal role in influencing the mead's color. The interaction between oak and mead can imbue the drink with rich amber tones, adding an aesthetically pleasing visual element. The degree of color change is influenced by the toast level of the oak and how long the mead is aged with it. Although this color enhancement might be subtler in meads with naturally darker hues, the nuanced amber infusion from the oak provides a visual richness that complements the drink's overall appeal.

Adding Structure Through Tannins

One of oak's important contributions to mead is introducing tannins. Tannins lend a fuller mouthfeel to the mead and a noticeably drier finish, creating a counterbalance to the inherent sweetness of the mead. This effect can make the mead appear drier than its actual residual sugar content might suggest, mirroring acidity's role in balancing sweetness. While moderate tannins enrich the mead, offering structural complexity, excessive tannins can lead to astringency. However, tannins are integral to aging, allowing mead to mature and evolve. Much like red wines, meads with a higher tannin content can age more gracefully, with the tannins mellowing over time to reveal a smoother, more refined drink.

Aging Mead in Barrels

Wooden Oak Barrel/Credit: Radoslaw Maciejewski (
www.Shutterstock.com)

This method infuses the mead with the intrinsic flavors of the wood but also introduces nuances from the barrel's previous contents and the subtle yet impactful effects of oxidation. Let's delve into the considerations and practices for oaking mead in barrels, whether new or used.

- **New Barrels:** Opting for a new barrel allows you to select from various wood species and toast levels, similar to choosing chips or spirals. New barrels directly contribute the wood's innate characteristics to the mead while permitting a controlled amount of oxygen to interact with the aging mead. This

160

interaction leads to a gentle softening of tannins and a refinement of the mead's fruity notes, enhancing its overall complexity.

- **Used Barrels:** For those seeking an extra layer of flavor, used barrels previously housing wine, whiskey, rum, cognac, or other spirits offer a special opportunity. These barrels can impart residual flavors from their past contents, adding a distinctive dimension to the mead. However, the size of commercial barrels, typically around 53 to 59 gallons, presents a challenge for the home mead-maker. Collaborating with fellow enthusiasts to fill a barrel is a practical solution that also fosters friendship and community!

It is generally advised to introduce mead to barrels after fermentation has concluded and the mead has clarified. This approach ensures that the barrel's influence is focused on aging rather than fermentation. During this phase, some evaporation is expected—affectionately known as the "angel's share"—which concentrates the mead's flavors and necessitates regular topping up to maintain the barrel's volume.

The size of the barrel plays a crucial role in the rate at which wood characteristics are imparted to the mead. Smaller barrels have a higher surface-to-volume ratio, accelerating the transfer of wood flavors and the absorption of notes from previous barrel contents. This rapid interaction requires vigilant monitoring to ensure the desired balance of flavors.

Maintaining cleanliness during barrel aging is paramount, especially when accessing the barrel for sampling or topping up. A simple yet effective practice is to sanitize the area around the bung with vodka before opening. Also, drilling a small hole to fit a stainless steel nail allows easy sampling without compromising the barrel's integrity. This setup enables mead-makers to draw samples cleanly, with the nail as a reusable, sanitary stopper.

Fine-Tuning Acidity

Adjusting acidity in mead is like painting with a palette of subtle flavors; it demands an artist's touch and a scientist's precision. There are different options for adjusting acidity.

Acid Additions

The careful addition of acidity can transform a mead, enhancing its profile with layers of complexity, brightness, and contrast. Understanding the unique properties of citric, malic, and tartaric acids is essential for any mead-maker aspiring to refine their craft and elevate their mead to new heights.

- **Citric Acid:** Infusing Brightness. Citric acid, known for its presence in citrus fruits, is the key to unlocking a refreshing vibrancy in mead. It introduces a crisp, clean brightness that can enhance the drink's appeal, especially in fruit-forward creations. The challenge lies in harnessing this acidity without tipping the balance towards an artificial sweetness. The secret? Incremental additions and consistent tasting. This approach ensures the mead remains true to its intended flavor, with just the right touch of zest.

- **Malic Acid:** Deepening Tartness. Malic acid, with its sharp and tangy character found in apples and pears, adds a compelling depth to mead. It's particularly effective in crafting meads that echo the qualities of traditional ciders or apple wines, offering a tartness that beautifully complements the sweetness of honey. To integrate malic acid successfully, introduce it gradually, mindful of its potent tartness. This method allows for a balanced incorporation, enriching the mead without overwhelming its delicate flavors.

- **Tartaric Acid:** Enhancing Sophistication. Tartaric acid stands out for its role in grapes and ability to impart a sophisticated, wine-like acidity to mead. Beyond its flavor contribution, tartaric acid offers benefits in stabilizing color and aiding fermentation, making it a multifaceted addition to the mead-maker's toolkit. Its use can improve the mead's aging potential, preserving its desired qualities. As with the other acids, adding tartaric acid should be measured and thoughtful, aiming to complement rather than dominate the mead's natural profile.

When making acid adjustments, the "less is more" principle cannot be overstated. Initiate this process by adding a modest quantity—consider beginning with a quarter teaspoon per gallon. This minimal approach allows the mead to assimilate the adjustment gently. After incorporating the initial acid dose, giving the mead time to marry with the new element is imperative. A few hours or, ideally, an entire day should elapse before conducting a taste test to evaluate the impact. This interval stabilizes the mead, enabling a more accurate assessment of the acidity's influence. Subsequent adjustments should only be contemplated after this reflective pause, safeguarding against the risk of over-acidification, a notoriously challenging condition to reverse.

The selection of acids plays a pivotal role in the crafting process, with food-grade, pure acid powders being the gold standard for their reliability and purity. Before adding to the larger batch, these acids should be pre-dissolved in a small volume of water or mead. This preparatory step is essential for ensuring a homogeneous distribution throughout the mead, eliminating the possibility of uneven acidity that could compromise the flavor balance.

Embracing Fruits for Their Natural Acidity

Incorporating fruits into mead presents a splendid opportunity to infuse natural acidity while enriching the mead with a complex bouquet of flavors. Opt for fruits known for their tartness, such as berries, cherries, and citrus, to introduce a harmonious blend of acidity and fruitiness. This technique adjusts the mead's acidity and layers the drink with a rich tapestry of flavors that complement the honey's natural sweetness.

The Creative Potential of Blending

In certain instances, the ideal acidity level may be attainable not through direct additions but through blending different mead batches. This method demands patience and the spirit of experimentation but can yield a mead that harmonizes the best qualities of each component, creating a "gestalt"—a sum greater than its parts.

Icing Your Mead

The process of icing, also known as eising, is a meticulous technique borrowed from the traditional practices of German Icebock production. This method concentrates the essence of mead by carefully freezing it post-fermentation, then selectively removing the liquid portion from the semi-frozen slush. The primary objective is to eliminate 10–20 percent of the volume as ice, which, interestingly, contains little flavor or alcohol, thereby enriching the remaining mead's alcohol content and intensifying its flavors.

To effectively employ the icing method, the mead must be frozen to a point where it achieves a slushy consistency. This enables the separation of the liquid mead from the ice, concentrating its characteristics. An alternative approach involves completely freezing the mead and then collecting the initial thaw liquid, which achieves a similar concentration effect. It's essential to recognize that the ice removed predominantly consists of water, with minimal loss of flavor or alcohol, resulting in a mead that is richer and more potent.

Understanding that icing is not exactly a remedy for subpar mead is critical. The process tends to remove water and some precipitating compounds, subtly enhancing the mead's overall profile. When done precisely, icing can refine the mead's aromas and smooth out any harshness, yielding a more aromatic and palatable product.

For mead enthusiasts interested in the icing technique, a small-scale experiment can be conducted using a sanitized plastic jug or bottle in the freezer until it reaches the desired slushy state. The mead should then be strained to collect the concentrated liquid. A notable consideration is that the freezing capabilities of a typical refrigerator's freezer compartment may not suffice for meads with higher alcohol content or significant sweetness levels. Additionally, it is advisable not to fill more than two-thirds of containers to mitigate the risk of overflow and mess due to expansion during freezing. While icing a sparkling mead may result in some loss of carbonation, a degree of effervescence will likely remain after the process.

Sulfites in Mead´s Long-Term Aging

Tracing its usage back to the ingenuity of the Romans, sulfur dioxide has been a cornerstone in preserving the integrity of wines. By burning sulfur within casks, they laid the foundation for what would become an enduring practice. Today, its legacy continues, with sulfur dioxide being integral in preventing oxidation, which is especially crucial for the nuanced profiles of mead.

Sulfur dioxide's versatility in food preservation is nothing short of remarkable. It functions as an antioxidant, forming sulfurous acid to avert oxygen's spoilage effects and as a redeemer of flavor, binding with aldehydes to banish any lingering harshness. It's a stalwart defender of color and balance, deterring the browning and oxidation that can mar a mead's aesthetic. Moreover, it keeps unwelcome microorganisms at bay, favoring yeast strains that withstand time and sulfite exposure.

Potassium metabisulfite is the preferred sentinel against spoilage for the dedicated home mead-maker. It doubles as a sanitizer, imparting a protective layer to equipment and the mead. When used judiciously, it can maintain the delicate bouquet of honey and secure the mead from the degradation that time can impart.

The Two Faces of Sulfur Dioxide

Sulfur dioxide comes with a caveat—it is as potent as it is effective. Its strong odor and bitter taste are the first hints that it must be used carefully. When introduced recklessly, it can provoke yeast into producing hydrogen sulfide, a compound far more offensive in aroma, capable of overshadowing the intricate flavors of mead. Moreover, an overdose may stun the yeast, disrupt fermentation, or inhibit necessary secondary fermentations, altering the intended character of the mead.

If sodium metabisulfite is part of your mead-making process, be mindful of its bitter edge. It can enhance undesirable flavors if not precisely measured and carefully integrated into the mead. Precision is key—aim for a harmonious balance where the mead's intended profile shines without the interference of unintended bitterness.

The most significant concern with sulfur dioxide is its capacity to provoke severe allergic responses. For individuals sensitive to sulfites, the reaction can be dangerous, even life-threatening. As a homebrewer, this risk should never be underestimated. Prioritize safety by being transparent about the presence of sulfites in your mead and understanding the health needs of those who will enjoy it.

Beyond the Basics: Exploring the Art of Aging Mead with Metabisulfite

For those of you who are just beginning to dip your toes into the vast ocean of mead-making, I'd like to offer a word of caution: adding metabisulfites is not a beginner's first step. This isn't to say that you won't get there, but rather, it's a gentle reminder that mead-making is more of a journey, not a race to the finish.

The beauty of mead-making lies in the learning curve, the experiments, and the gradual building of expertise. When adjusting SO_2 levels using potassium metabisulfite after fermentation, the process requires a keen understanding of chemistry. This step is crucial for those aiming to perfect their craft, ensuring the mead ages beautifully, and developing the rich flavors and aromas that make this beverage so revered.

For you spirited souls eager to delve into the complexities of this process, don't fret! While I've chosen not to overwhelm you with the technicalities here, I encourage your thirst for knowledge. More detailed insights and guidance await those ready to take the next step.

By scanning the QR code below, you'll unlock a treasure trove of information meticulously designed to guide you through the more advanced stages of mead-making at a pace that suits you.

Your adventure into the realm of mead-making is taking an exciting turn, and it's time to look into the nuances of aging your creation to perfection. The use of sulfites, specifically SO_2, is a cornerstone in crafting a mead that not only stands the test of time but also develops into a beverage bursting with depth and character.

Whether you're well-versed in the traditions of mead-making or just beginning to explore this fascinating craft, the way ahead is filled with discoveries.

I have handpicked a couple of resources that shine a light on how to harness the power of sulfites to enhance your mead. These websites are treasure troves of knowledge, offering insights that bridge the gap between ancient practices and modern techniques.

Or you can click the link: www.bit.ly/SO2_for_agin_gmead

Carbonation

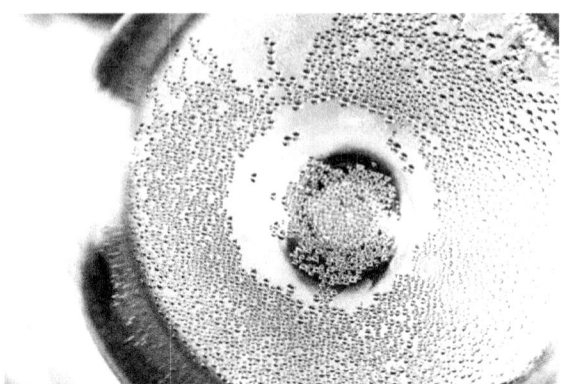

Carbonation/Credit: Don Pablo (www.Shutterstock.c om)

Excited about giving your mead some fizz? You're in for a treat! Carbonating mead is incredibly rewarding and more straightforward than you might think!

Let's start at the end—bottle conditioning. This is where the carbonation magic happens, but of course, after you've carefully gone through the primary and secondary fermentation processes.

Choosing the right type of mead for carbonation is crucial. Think of it as selecting the perfect candidate for a sparkling transformation, like determining which wines to turn into sparkling varieties or deciding which beers to carbonate.

Not every mead is suited for carbonation. For example, the rich and heavy profiles of port, Shiraz, or most pinot noirs, much like dense stouts, don't mesh well with bubbles. They're best enjoyed in their still form. On the other hand, meads with lighter body and less sweetness are prime candidates for carbonation, ready to be transformed into something quite extraordinary.

Now, for the surprisingly straightforward part: once you've completed secondary fermentation, it's time for bottle conditioning. And here's where a simple addition plays a pivotal role—more honey! This isn't about alcohol content; it's about using the natural sugars to initiate a *third* round of fermentation right in the bottle. The yeast jumps back into action, consuming the sugars and releasing carbon dioxide as a byproduct. But unlike during the first stages of fermentation, where gas escapes through the airlock, the sealed bottle traps the CO_2, dissolving it into the mead and creating those delightful bubbles.

This approach is elegantly simple. The process hinges on the closed system of the bottle, which ensures the carbon dioxide has no escape, leading to beautifully carbonated mead. It's a testament to the joy of mead-making, where a touch of honey and a little yeast can create a bubbling sensation that elevates your mead to new heights!

Choosing Your Bottles and Seal

The first step is often overlooked yet crucial: selecting the appropriate bottles. It's tempting to think any wine bottle will suffice for your carbonated concoction, but this choice can make or break your effort. Regular wine or mead bottles might not withstand the pressure from carbonation, risking a disastrous cleanup of mead and shattered glass. Imagine explaining that to your partner!

For a successful sparkling mead, your bottles need to be as resilient as they are elegant. Champagne bottles are the gold standard in homebrewing circles for their strength against effervescent pressure. Whether you opt for brand-new or lovingly reused champagne bottles, their durability is non-negotiable. As an alternative, consider the robust 22-ounce "bomber" beer bottles, or 750 ml beer bottles, which are also conditioned to withstand the rigors of carbonation.

Ensuring a secure seal is paramount once you've got the suitable bottles. The unique champagne cork, designed to absorb some carbon dioxide and expand, creates an optimal seal with a wire cap. This dynamic duo is essential in keeping the sparkling magic safely contained within your mead.

From Fermentation to Effervescence

Effervescence begins once your mead has fully fermented, utilizing all available sugars. Imagine starting with a specific gravity of 1.060 and ending at 1.000. This is where dormant yeast springs back into action, relying on you to introduce a new source of fermentable sugar for priming—the key to unlocking carbonation within the bottle. The importance of *precision* at this stage cannot be overstated; using a bottle carbonation calculator is crucial to avoid the dreaded over-carbonation or, in worst-case scenarios, explosive "bottle bombs."

After carefully adding and mixing your chosen priming sugar, store the bottles at room temperature to let the yeast work its magic, typically for 1 to 2 weeks. A pro tip: chill a bottle after a week to test the carbonation level, giving you a sneak peek into the fruits of your labor.

Choosing Your Priming Sugar

The selection of priming sugar offers a palette of flavors and carbonation levels, from corn sugar and sucrose to more unique options, like maple syrup or honey. Each brings character to your mead, influencing the final bouquet and bubble intensity.

Employing a priming sugar calculator, such as the one provided by Northern Brewer, is an invaluable step in this process. It helps tailor the carbonation to your preferences, ensuring a perfect fizz without the risk of overdoing it.

You can access the calculator at the QR code below:

Or you can click the link:
https://bit.ly/prime_sugar_calculator

Blending your Mead

Blending means combining different meads, or even integrating other beverages, thus offering a range of sources for the mead-maker to correct, enhance, or innovate their brews.

Blending serves various roles in the mead-making process:

- **Achieving Balance:** Merging a sweet mead with one that's dry or has a touch of acidity can create a beautifully balanced beverage. The sweetness can mellow the acidity or dryness, making the mead palatable without overshadowing its innate flavors.

- **Softening Overpowering Elements:** A too-bold mead in fruit, spice, or oak can be harmonized by introducing it to a traditional mead with a milder flavor, perhaps highlighted by a distinctive honey varietal.

- **Leveraging Excess Sweetness:** An overly sweet mead finds its place as a valuable tool for back-sweetening other batches, providing a clean, controlled sweetness.

- **Creating Depth:** Inspired by the tradition of blending old and new beers, combining aged mead with a fresher batch can yield a mead with layered complexity, offering the best of both worlds.

- **Innovative Flavors:** Experimenting with blends of different mead styles or even incorporating wines, beers, or ciders can lead to groundbreaking flavors and styles, opening up a world of experimentation.

The Blending Process

First, one must understand each component's flavor profile to begin the blending journey. I know what each mead brings and envision how they might complement or contrast.

The blending story unfolds as follows:

- **Setting the Stage with Vision and Understanding:** Every great blend starts with a spark of inspiration. This could be to meld the bright acidity of one mead with the luscious sweetness of another or perhaps to marry various batches for a result greater than its parts. The initial step is to paint a mental picture of the desired outcome, considering the flavor, aroma, and mouthfeel you're after. This vision guides every subsequent decision in the blending process.

- **Small-Scale Experimentation:** Mix small samples from each source in a separate container. This step allows you to trial your vision without risking your entire batch.

- **Choosing Your Palette:** Selecting the components for your blend is like choosing colors for a canvas. This stage involves tasting and evaluating potential candidates for their potential synergy, not just their merits. It's a moment to assess the characteristics of each mead, pondering how they might complement or contrast each other to achieve the envisioned blend. The deliberate and intuitive selection lays the groundwork for the creative journey ahead.

- **Experimentation on a Micro Scale:** The next step is experimenting with small-scale blends with the chosen components. Mixing small amounts in various ratios allows you to explore different combinations, observing how they interact and affect the overall profile. This phase allows for fearless experimentation, where each trial brings you closer to the perfect blend without risking large volumes of mead.

- **Refining Through Tasting:** Each sample blend is carefully evaluated for alignment with the initial vision. Adjustments are made in small increments, with the blend evolving with each iteration. This step is guided by a critical palate and personal preference, aiming to strike the perfect balance that resonates with the intended flavor profile.

- **Scaling Up With Precision:** Once the ideal blend is discovered on a small scale, the challenge becomes scaling it up. This requires careful calculation to adjust the proportions accurately for a larger volume. The process is meticulous, with an understanding that the dynamics of flavor may subtly shift with volume. Additional tastings ensure that the larger batch maintains the integrity and harmony of the small-scale success.

- **The Final Tasting:** Before the blend is finalized, one last comprehensive tasting confirms that the larger batch captures the desired characteristics. This final step is both a validation of the blending process and an opportunity for any last-minute fine-tuning.

- **Documenting the Process:** An often overlooked but essential aspect of blending is documentation. Keeping detailed records of the blending process—from the initial ratios and adjustments to the final recipe—ensures that successful blends can be replicated and that each experiment, regardless of its outcome, contributes to your broader knowledge and skill.

CHAPTER 6
Preparing Meads With Additional Ingredients

T his chapter builds upon earlier content in the book and unveils the exciting possibilities that additional ingredients bring to your mead! From the rich flavors of fruits to the subtle nuances of spices and herbs and the robust character of malt, we'll explore how these elements can transform your mead into a complex, multi-layered beverage. I will provide practical advice on incorporating these ingredients seamlessly into your brews, ensuring they complement rather than overwhelm the delicate balance of flavors.

Fruits(Melomels)

*Fruit and Honey/Credit: Maglara (www.Shutterstock
.com)*

The joy of melomel crafting lies in the unexpected! Remember, the mead you'll brew may surprise you, taking on a life distinct from the fresh fruit it once was. The process shares its roots with traditional mead-making, but here's where you'll take a thoughtful detour. It's about finding the proper harmony between your honey, the fruit's boldness, and the nuances of fermentation. Consider the fruit's journey from vine or tree to bottle — whole, crushed, or juiced, it's all about the essence it brings. Even the water takes a backseat when fruit juices step in to fill the role, like apple juice in a cyser, creating a beautiful base for your mead.

Honey Selection

Don't fret over varietals when choosing your honey for that bold melomel. Often, the fruit's vibrant character is the star of the show. But, pause for a moment with the subtler fruits — that's where a carefully chosen orange blossom or tupelo honey can elevate your mead to new heights. Wildflower honey is often a great pick with its delightful yet economical allure. Just be mindful, as its profile shifts with the seasons, and avoid batches that carry too much of nature's wild side, like the aromatic dandelion or the intense basswood.

The Fruit of Your Labors

Picking your fruit is a bit like choosing the theme for a party—what do you love? What excites your senses? It will likely sing a chorus in your mead if it's joyous to eat. Berries bring a burst of joy: strawberries, raspberries, and blackberries, each with their own fanfare. Stone fruits whisper of summer's sweetness with their soft, juicy flesh. Don't hesitate to mix and match — your taste buds are your best guide. And when it comes to crafting Pyments or cysers, remember to play with the acidity and sweetness. High-acid grapes and tangy apple cider blend marvelously with the rich sweetness of honey, creating a luscious blend of flavors that truly belong together in your glass.

Fruit Hunting

When searching for the perfect fruits to infuse your melomels, bypass the supermarket aisles. Those fruits are picked for their looks and longevity, not for the lush flavors and aromas that your mead deserves. Think about it—your future self won't recall the extra dollars spent on quality fruits, but you'll undoubtedly taste the difference with every sip of that sumptuous melomel.

Here's the game plan: put on your detective hat and track down those heirloom, flavor-packed varieties. Before mega grocery stores became the norm, fruit was bred for its rich taste, sugary sweetness, and juicy bite. Thankfully, some dedicated farmers still honor these traditions, and their produce might even be waiting for you at a local nursery, ready to take root in your backyard. Growing your own means you can pamper your plants and pick the fruits at their peak—a surefire way to elevate the quality of your mead.

Start your quest at the local farmers' market or flip through your phone book. Connect with growers who can tell you about their fruit varieties; they're the good folks who will be thrilled to guide a fellow fruit lover. Supporting these passionate orchardists enriches your mead and helps sustain their craft.

Don't let the myth that top-tier fruits are unattainable unless you grow them yourself deter you. With a keen eye, you can snag exquisite fruits almost anywhere. Farm markets are treasure troves—if you can tell a grower from a reseller. Remember, the best flavors often don't come in the prettiest packages, and patience can lead you to these hidden gems.

Lastly, if you're adventurous, consider visiting a produce terminal. They're mainly for the trade, yes, and they keep odd hours, but the variety is mind-blowing. You might walk away with a haul of premium fruits at a price that's very kind to your wallet.

Exploring the addition of fruit to mead is a journey where the path you choose — the fruits and their incorporation into your mead — can lead to wonderfully diverse outcomes. As we venture forth into this discussion, it's important to remember that the best method is the one that *works for you.* You've struck gold if it delivers the desired results and you find the process manageable. Let's not get tangled in a web of rigid rules that drains the joy from our craft!

Mead-making with fruit can take many forms, and while my preference leans towards fresh, whole fruits for their vibrant quality, don't overlook other forms like juices, concentrates, or even fruit liqueurs for their ease and consistency.

Fresh Fruits

The allure of fresh fruits is undeniable. Imagine wandering through a "you-pick" farm or meandering among the bustling stalls of a local farmers' market; this is where the quest for the perfect fruit begins. The freshness of your selection is paramount—not just for the vibrant flavors they promise but also for the purity they must ensure. A simple dip in cool water can reveal the true gems in your harvest, separating the ripe and ready from the unwanted. A gentle rinse will suffice for those tender fruits, like raspberries, that bruise at the slightest touch. And if you're looking to lock in those flavors for the perfect brew moment, freezing your fresh finds can be a game-changer. Remember, this trick doesn't apply to all—apples and citrus prefer the immediacy of freshness; their zest, and tang best captured live.

Frozen Fruits

Sometimes, the stars don't align; your desired fruit is out of season, or your brewing schedule is already at capacity. This is where frozen fruits shine as your faithful stand-ins. Whether scouring the frozen section of your local grocery store or hitting up the vast selections of a big-box retailer, the right frozen fruit can keep your melomel aspirations on track.

It's important to debunk the myth now: freezing does not *cleanse* the fruit. It merely hits the pause button on decay and dormancy until warmth breathes life back into those microbes. Yet, this frosty detour can be a boon for brewers, softening the fruits to yield their full spectrum of flavors to the fermenting frenzy. That said, vigilance against freezer burn is crucial; this ailment can taint your fruit with unwelcome flavors that no brewing wizardry can erase. Opting for quality freezer bags and minimizing their icy exile are your best defenses. A sulfite spray before freezing can work wonders for those looking to fortify their fruit against the ravages of oxidation. Even in commercially frozen fruits, beware—the slightest breach in their packaging can invite in the elements you seek to avoid, leaving your melomel dreams frozen in time!

Juices and Concentrates

Juices and concentrates are a boon for mead-makers, offering a straightforward path to infuse fruit flavors. The key lies in the selection process: aim for products that stay true to the essence of the fruit, free from added sugars or artificial enhancements. This purity means that the fruit's natural sweetness and character become the stars of your mead. Beware of preservatives such as potassium sorbate, which can bring fermentation to a standstill, and consider pasteurized options, though they may slightly alter the fruit's vibrancy. For those leaning towards concentrates, look for ones that blend smoothly, leaving behind minimal sediment, making your mead-brewing process as efficient as possible.

Purées

Purées open up a convenient avenue for incorporating fruit into mead, especially in terms of storage and preservation. These aseptic options in cans or durable Mylar bags promise a long shelf life and a ready-to-use fruit base. However, it's worth noting that purées might offer a softer echo of the fruit's full spectrum of flavors when compared to their fresh or frozen counterparts. Everything in the purée, once introduced to the fermenter, will eventually settle, a reminder of the transformation occurring within. For those drawn to canned fruit purées, selecting those preserved in natural juices or water, and free from any fermentation-inhibiting preservatives, is critical to maintaining the integrity of your fermentation process.

Wine and Fruit Kits

For mead-makers seeking variety and consistency, the local homebrew supply store is a treasure chest filled with wine and fruit ingredient kits. These kits aren't limited to traditional grape varieties; they span a range of fruit flavors, from the tartness of raspberries to marionberries' unique taste. Grape wine kits, handy for crafting pyments, typically include ample juice to set the foundation of your mead. When exploring fruit wine kits, the emphasis should be on those that boast a composition of pure fruit—avoiding any with added sugars. This approach ensures the natural flavors and qualities of the fruit are the highlights of your mead.

Fine-Tuning Fruit Quantities

The quantity of fruit you decide to infuse into your mead can transform its character from subtly fragrant to richly vibrant. This balance hinges on several factors: the type of fruit, the desired sweetness level, and personal taste preferences. For those venturing into sweeter meads, a generous addition of fruit will achieve a harmonious balance. Typically, berries and stone fruits demand a volume of at least three pounds per gallon to impart their essence, with many recipes leaning towards 3.5 to 4 pounds for a more pronounced flavor.

However, when it comes to fruits with a bolder presence, like blackcurrants, less is more—about 2 pounds per gallon should suffice. Starting with 3 pounds per gallon offers a solid foundation for most fruits, allowing you to adjust according to taste. For those intrigued by layers of complexity, consider adding fruit in stages throughout the fermentation process.

The interplay between the sweetness of honey and the fruit's acidity balances the mead and accentuates the fruit's authentic character. For enthusiasts favoring a drier melomel, scaling back to 1 to 1.5 pounds of fruit per gallon—and keeping alcohol content under 10% ABV—can mitigate any potential harshness in the finish.

Varieties such as pyments and cysers traditionally replace water with raw juice in the mead mixture. A standard 5-gallon batch might blend 1 to 1.5 gallons of honey with 3.5 to 4 gallons of specific grape juice or cider. It's crucial to monitor the pH of the juice; if it is too low, consider adjusting it upwards or diluting it with water for a smoother fermentation.

Timing the Addition of Fruit in Melomel-Making

Deciding just *when* to add fruit to melomel is a subjective choice, sparking lively discussions among mead-makers. One prevalent view is that introducing fruit during secondary fermentation captures its true character and aroma most effectively. While beneficial for flavor infusion, this approach requires patience; the fruit must be immersed for several weeks, or months, to release its full potential into the mead. Additionally, this technique calls for an extended aging period to ensure the fruit and honey flavors meld seamlessly. It's worth noting that this method might kickstart fermentation, again, due to the natural sugars in the fruit, especially if the primary fermentation concluded because the alcohol level reached the yeast's tolerance.

An intriguing alternative is to ferment the fruit and honey separately and blend them after both fermentations have finished. This method shines in pyments or cysers, blending wine or hard cider with mead to achieve the desired flavor profile. It offers the flexibility to adjust the fruit-to-mead ratio and experiment with different fruit sources or grape varieties. Although this technique allows for creative blending, it may slow the integration of flavors and result in distinct fermentation characteristics compared to simultaneous fermentation.

For those aiming to maximize fruit flavor, adding raw fruit after the mead has been stabilized post-fermentation is another option. This approach ensures the freshest fruit flavor but can also introduce a sweetness reminiscent of unfermented fruit, alongside a slight risk of bacterial contamination. However, since the mead already contains alcohol, the risk of spoilage is minimized. This method is typically employed to fine-tune the fresh fruit aspect of the mead, rather than as the primary means of fruit incorporation.

Incorporating fruit at the outset of primary fermentation is yet *another* strategy, demanding meticulous sanitation to prevent microbial invasion. Pre-treating the fruit by washing, freezing, and thawing reduces bacteria and enhances juice extraction during fermentation. Introducing fruit in primary fermentation enriches the mead with essential nutrients from the fruit and aids in pH balance, offering a quicker fermentation cycle. Although some of the most volatile aroma compounds may dissipate, they generally evaporate during aging. For a layered fruit experience, adding a pound of fruit per gallon during secondary fermentation can introduce an additional layer of complexity, enriching the melomel without overshadowing the base flavors developed during primary fermentation.

Fruit Preparation

Imagine strolling through a local market or garden, choosing the best fruits for your mead. Being picky here is crucial—only the freshest, most vibrant fruits make the cut. Anything less, anything you'd *hesitate* to eat, doesn't deserve a place in your mead!

Once you've gathered your fruit, it's time for prep work. This isn't just about cleaning; it's about enhancing flavor. You'll want to remove pits for fruits like peaches and plums (except for cherries, where the pits add a unique character, provided they're removed on time). Then, give your fruits a good wash and freeze them solid (if you use fresh fruits). Freezing isn't just a storage hack; it's a flavor enhancer, breaking down cell walls to release the juicy essence of your chosen fruits.

Adjusting pH: The Balancing Act for Perfect Fermentation

A healthy fermentation process hinges on maintaining the correct pH. The pH naturally dips as fermentation progresses, but it's a balancing act. Too low, and you risk stalling the process, leaving your mead in limbo.

Thanks to their potassium content, fruits are your allies in maintaining pH balance. However, ensuring your must starts at a pH of 4.0 sets the stage for success. Potassium carbonate is an excellent choice for this adjustment. It doesn't just tweak pH levels; it also supports yeast health with a potassium boost.

But, as with all great things, moderation is key. Overuse of carbonate can tip the scales, affecting the must's total acidity and, ultimately, the taste of your mead. It's about finding that sweet spot, where everything harmonizes to create a mead that's not merely good but memorable.

Adding Fruit to the Fermenter

Plastic fermenters are your allies in this endeavor, offering unparalleled flexibility. Envision wrapping your chosen fruits in a muslin bag or a nylon stocking. This isn't just about keeping things tidy; it's a strategy for infusing your mead with fruit essences without hassle. When it's time, you can lift the bag and enrich your compost with the spent fruit, returning it to the earth.

For a smooth experience, a cheesecloth or mesh bag can be a game-changer. Picture placing your vibrant fruits into this bag, sealing it with a zip-tie that's been sterilized to perfection. This method isn't just about efficiency; it's about preserving the integrity of your mead, making racking a breeze and ensuring not a drop of your precious creation is wasted unnecessarily. If you go bagless, a sanitized kitchen strainer will be your best friend, gently removing fruit while safeguarding your mead's clarity.

Glass carboys hold their charm and challenges. Without the convenience of a bag, preparation becomes critical. Set your fermenter aside for siphoning a few days before you plan to rack. This patience pays off, allowing your mead to settle, clearing your path to a smoother racking process.

Fruit-Specific Insights

Every fruit tells a story! Strawberries and raspberries, for instance, might sprinkle your mead with seeds and bits of pulp, a testament to their natural essence. While some might see this as a challenge, it's also an opportunity to embrace their natural texture. Yet, if clarity is your goal, patience and multiple rackings can help, though each step might claim a bit of your mead. It's a dance of compromise and decision-making, where you decide the balance between perfection and yield.

An extra layer of patience is required for fruits like peaches, plums, and pears, which tend to surrender to the fermentation process, becoming mushy.

Managing the Cap

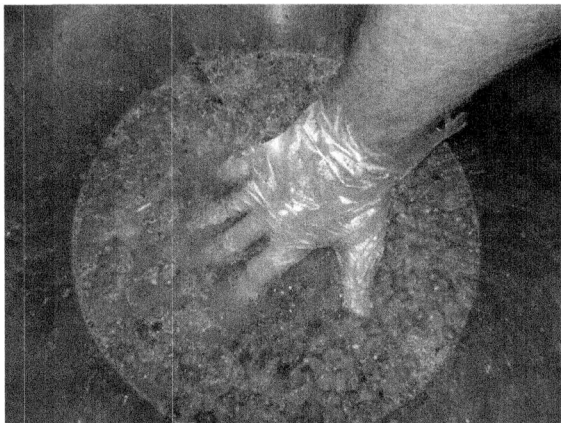

Punching Down the Cap/Credit: BDoss928 (www.Sh
utterstock.com)

This layer of fruit that floats to the top, cradled by bubbles of CO_2, is more than just a visual marker of fermentation—it's a critical factor in the ultimate taste, aroma, and quality of your mead.

Why the fuss over the cap? It's simple: the cap's management is directly linked to capturing those lush fruit flavors, ensuring yeast thrives without stress, avoiding stalled fermentations, and sidestepping any unwelcome off-flavors. Think of the cap as a gatekeeper to the quality of your mead; how you manage it will make all the difference.

Punching down the cap isn't just a task—it's an essential ritual in the mead-making process. This act releases trapped CO_2, which, in excess, could harm your yeast and muddle the fruit's natural flavors. Moreover, it helps keep the fermentation's temperature in check, preventing the heat beneath the cap from reaching levels that could endanger the yeast's well-being.

But it's not only about avoiding the bad; it's also about promoting the good. Aerating the must by punching down ensures that yeast cells get enough oxygen to produce sterols for their cell walls, enhancing their strength and fermentation capacity.

Allowing the cap to dry out invites many potential issues, including the risk of spoilage organisms taking over if oxygen finds its way in. These unwanted guests can compete with yeast for nutrients, stressing them and potentially leading to a less vibrant, less clean-tasting mead. By mixing the fruit back into the must, you're not just preventing these issues but also ensuring that every ounce of fruit character—color, tannin, aroma, flavor—finds its way into your mead.

With the highest concentration of yeast living in and just below the cap, evenly distributing this layer throughout the must can significantly impact the fermentation's health and vigor. This process ensures that yeast, sugar, and temperature are evenly spread, fostering a uniform and efficient fermentation process and reducing the risk of an uneven or stuck fermentation.

Crafting a Melomel: The Basic Process

Here we cover the essential steps for incorporating fruit into your mead, sticking to the basics without venturing into the more advanced techniques like back-sweetening, stabilization, and clarification (discussed earlier in the book).

If you're looking to add a fruity twist to your mead, you'll find that the process largely mirrors the one we've already explored, with a few key additions to accommodate the fruit.

The first step in your melomel adventure involves preparing your honey. Gently warming the honey containers in hot water will make the honey easier to work with, allowing for a smoother blend with water. If you're using frozen fruit, allow it to thaw so it mashes more easily and integrates better with the honey. Remember, "cleanliness is next to godliness" in mead-making. Ensuring that all your equipment is sanitized cannot be overstated—it's the foundation of a successful fermentation.

Once your ingredients are ready, it's time to mix the must. Add your prepped fruit to the fermenting pail. The ideal temperature for your must is a cozy 65 to 70 °F. If your fruit has lowered the temperature, gently warming the water can help nudge it back into the sweet spot. Now, introduce the honey and water to the pail, using a drill-mounted wine degasser to ensure the honey dissolves fully. After fully integrating the honey, give the must a good stir to aerate it, setting the stage for a healthy fermentation. It's crucial here to avoid heating the must, as we want to preserve the natural integrity of the honey and fruit.

With the stage set, it's time to introduce the yeast to its new environment. Rehydrate your yeast following the packet's instructions, adding a rehydration nutrient like Go-Ferm to give it a fighting chance in its new sugary home. Once rehydrated, pitch the yeast into the must, add your first dose of nutrients, and mix well. You should see signs of fermentation within 24 hours, marking the beginning of your mead's transformation. Managing the fruit cap and following your nutrient schedule are your next steps. A successful fermentation typically wraps up in two to four weeks, but patience is critical—rushing will only harm the process.

After allowing the mead to ferment in the primary vessel for about four weeks, it's time to transfer it to a secondary container for clarification. If clarity eludes your mead after a month, consider employing a two-stage clarifier like Super-Kleer. Patience during this stage is crucial; ensure that fermentation has completely stopped and your mead is crystal clear before bottling.

Understanding Fruit Water Content in Melomel Making

Fruits, lush and bursting with flavor, are predominantly made up of water, typically around 85%, and contain relatively modest sugar levels, averaging just below 10%. This composition is crucial to consider as it significantly influences the specific gravity and final sweetness of your melomel.

Let's break it down with a practical example: imagine you decide to enrich your mead with 15 pounds of raspberries. Since raspberries are 85.75% water, you can add roughly 1.54 gallons to your mead. This isn't just a splash of water; it's a considerable volume that can dilute your must. Moreover, the sugar contribution from these raspberries, at a 4.42% sugar content, is relatively minimal, about 0.66 pounds. This dilution and modest sugar addition require careful consideration to achieve the desired final product.

Understanding that fruits release their water content gradually into the must means you can't rely solely on your hydrometer readings right after fruit addition to gauge the future specific gravity of your mead accurately. For instance, adding 15 pounds of raspberries to a 5-gallon batch with an initial gravity of 1.130 could lower your starting gravity to 1.092 due to the water content, potentially shifting the mead from a medium-sweet profile to a drier finish. This insight underscores the importance of calculating the water and sugar content of the fruit accurately, ensuring you steer your melomel towards the desired taste and strength. Here's a handy formula to keep in your toolkit:

$$Gallons\ of\ Water = \frac{\frac{Fruit\ Weight\ in\ Pounds * Percentage\ of\ Water\ in\ Fruit}{100}}{8.33}$$

With 8.33 being the weight of one gallon of water in pounds. Water's weight at standard temperature and pressure is approximately 8.33 pounds per gallon.

The Role of Pectin

Pectin, a natural substance found in most fruits, is renowned for its ability to gel, a property that's essential in making jams and jellies. However, when it comes to making melomel, pectin poses a challenge: it can cause a persistent haze that is stubbornly difficult to clear. The issue becomes more pronounced when the fruit is heated, as the warmth initiates the gelling process of pectin, increasing the risk of haze in your mead.

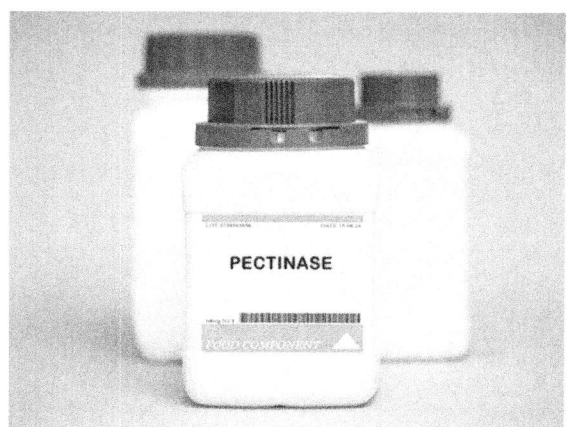

*Pectinase/Credit: luchschenF (www.Shutterstock.co
m)*

Managing pectin is critical for those looking forward to a clear, visually appealing melomel. Introducing pectinase enzyme into the must alongside the fruit emerges as a proactive strategy to counteract the haze. Pectinase breaks down the pectin, mitigating its gelling effect and thus preventing the haze from forming. If clarity issues persist, adding pectinase later can help achieve the desired transparency. Given the variety of pectinase products on the market, adhering to the specific usage instructions provided on the packaging is crucial for optimal results.

This approach not only preserves the alluring clarity of melomel but also ensures a smoother filtration process, should you choose to filter your mead. Pectin, without intervention, can clog filters, complicating the clarification process.

Below is a table with an overview of the water and sugar content and pectin level of most fruits often used for preparing melomels.

Fruit	Water %	Sugar %	Pectin Level	Fruit	Water %	Sugar %	Pectin Level
Apple Juice	88.24	9.62	High	Apricots	86.35	9.24	Low
Bananas	74.91	12.23	N/A	Blackberries	88.15	4.88	Low
Blueberries	84.21	9.96	Medium	Cantaloupe	90.15	7.86	Low
Cherries (Sour)	86.12	8.49	Medium	Cherries (Sweet)	82.25	12.82	Medium
Cranberry Juice	87.13	12.1	High	Black Currants	81.96	N/A	High
Red/White Currants	83.95	N/A	High	Elderberries	79.8	N/A	Low
Dried Figs	30.05	47.92	N/A	Raw Figs	79.11	16.26	N/A
Gooseberries	87.87	N/A	High	Grapefruit Juice	90	9.1	N/A
Muscadine Grapes	84.29	N/A	N/A	Guavas	80.8	8.92	High
Honeydew	89.82	8.12	Low	Kiwi	83.07	8.99	N/A
Lemon Juice	92.31	2.52	High	Lime Juice	90.79	1.69	High
Mangoes	83.46	13.66	N/A	Mulberries	87.68	8.1	N/A
Nectarines	87.59	7.89	Low	Orange Juice	88.3	8.4	High
Papayas	88.06	7.82	N/A	Peaches	88.87	8.39	Low
Pineapple Juice	86	12.82	High	Raw Pineapples	86	9.85	Low
Plums	87.23	9.92	High	Pomegranate Juice	85.95	12.65	Low
Pomegranates	77.93	13.67	Low	Prickly Pears	87.55	N/A	N/A
Raspberries	85.75	14.42	Medium	Rhubarb	93.61	1.1	Medium
Strawberries	90.95	4.89	Low	Tangerine Juice	88.9	9.9	N/A
Watermelon	91.45	6.2	Low				

Other Considerations

Here are some tips that will elevate your melomel to the next level:

- First and foremost, when working with citrus or other pithy fruits, be cautious of the white layer—the pith—between the peel and the fruit. This part can impart a bitter taste to your mead, which might not always be welcome. However, exceptions exist where a balance between sweetness and the pith's bitterness is desired.

- Dried fruits can be a fantastic addition to your mead, but checking for sulphites is vital. These compounds are often added as preservatives but can inhibit yeast activity, stalling fermentation. Processed dried fruits are fine for snacking, they're not really suitable for your brew!

- If your chosen fruits have pits, make sure to remove them. Pits can introduce unwanted flavors and potentially harmful compounds into your mead. It's a simple step that can prevent complex issues down the line.

- Don't hesitate to experiment with adding fruit at different stages of fermentation. While primary fermentation can transform the fruit's flavors, adding fresh fruit to the secondary can reintroduce and enhance subtle nuances. This technique allows for a more layered and complex fruit profile in your final product.

- Lastly, be mindful of fruits high in oils, such as olives and avocados. While not traditional choices for melomel, if you're feeling adventurous, be aware that their oils can introduce unique challenges, including affecting clarity and mouthfeel.

Most Common Fruits Used in Mead-Making

This segment is dedicated to shedding light on some of the most beloved fruits in the mead-maker's palette. We'll focus on the essence of each fruit, uncovering the regions where it thrives, its culinary uses beyond the raw form, and, most importantly, the unique attributes it lends to mead.

While this discussion will focus on these fruits' raw, unfermented characteristics, it's important to note that fermentation can transform their sweetness and flavor profiles, sometimes making them difficult to identify in the finished mead!

The diversity of fruit used in mead is vast, with each type offering its distinctive taste and aromatic profile. It's essential to recognize that the essence of some fruits is so unique that describing them without referring to the fruit itself is nearly impossible—their color, aroma, and taste are definitive. Fruits also vary widely in sweetness, acidity, tannins, and flavor, influencing the mead's color and overall sensory experience in remarkable ways.

As we embark on a brief alphabetical tour of fruits commonly used in mead, remember that this list is merely a starting point. The true limit to a melomel's fruit composition is the mead-maker's imagination and willingness to experiment.

Apples

Think of apples not just as fruit but as a key ingredient that brings a spectrum of flavors to your mead. With a pH between 3.30 and 4.00, apples offer the perfect balance of tartness, essential for that bright, crisp note in your brew.

Imagine the vast array of over 7,500 apple varieties worldwide, with the United States alone boasting 2,500. When selecting apples for your mead, consider the endnote you aim for. Each variety has its own identity that can define the character of your mead. Will you choose a crabapple's bold bitterness or the honeyed tones of a golden delicious? Your choices in apple varieties are as influential as the yeast you pitch, or the water you use.

But apples bring *more* than just flavor. They're also about visual appeal—imparting a beautiful range of colors from soft straw to rich gold. That's the magic of mead-making!

We are engaging all the senses. And let's not forget the aroma. The right apple can transform your mead's aroma into a fragrant bouquet reminiscent of a fresh orchard at peak harvest.

Remember, the apple's contribution to your mead commands delicate balance. Too much acidity and your mead could end up too sharp; too little might lack that refreshing zest that mead drinkers love. It's about finding that *sweet spot*, where every sip is a harmonious blend of flavor, aroma, and color.

Apricots

Adding apricots to mead brings a touch of East Asian tradition to the craft, a nod to the fruit's native soil. When ripe and ready, the fruit itself is a delight, boasting a spectrum of color from a warm, sun-kissed yellow to a deep, inviting orange, occasionally dressed in a subtle reddish tinge. Its smooth, almost hairless skin is as pleasant to the touch as its flavor is to the taste.

Apricots contribute a fermentation value of six points per pound, with their natural pH ranging between 3.30 and 4.80. This lends a versatile acidity to the mead, allowing for a good balance between sweet and tart. The infusion of apricots doesn't just stop at taste; it sweeps into the aroma, imparting a robust and peach-like scent that promises a mead with a personality as rich as its color. This stone fruit, characterized by its single, almond-like seed, is often celebrated in pies and juices, and in mead it offers a golden-orange hue that is visually striking. When sipped, apricot mead is refreshingly mild.

Blackberries

Blackberries bring a touch of the wild to the art of mead-making, with each berry's deep purple hinting at the mystery of the forests where they're often found. These fruits, an assembly of small drupelets clustered together on brambles, offer a distinctive tartness and a rich color that will deeply influence the profile of a mead. Originating from various parts of the globe, including Oregon's fertile lands and the varied terrains of Asia, Europe, and the Americas, blackberries have a rich cosmopolitan heritage.

Integrating blackberries into mead balances acidity and tannins, with each pound of fruit contributing approximately three fermentation points (0.003 specific gravity points) and a pH value between 3.90 and 4.50. The tannins they provide lend a structured mouthfeel to the mead, while their vibrant hue promises a visual feast. In culinary uses, blackberries are as versatile as they are in mead-making. They are celebrated in everything from the simplicity of fresh fruit to the decadence of pies and the concentrated burst of flavor in juices. Adding blackberries to mead isn't only about infusing great flavor—it's about crafting a beverage experience that's both rooted in tradition and vibrant with innovation.

Blackcurrants

Blackcurrants, the small yet powerful fruits that hail from the high altitudes of Tibet, are a bold choice for mead-makers seeking to imbue their concoctions with depth and intensity. These berries, known for their deep purple-to-black color, contribute a rich, vinous character to mead, offering a robust and intense flavour profile with hints of earthiness that some describe as slightly "catty".

Incorporating blackcurrants into mead enhances its hue and introduces six fermentation points per pound, with an acidity level between pH 2.6 and 3.1. This acidity and the fruit's significant tannin content lay the foundation for mead with considerable structure and complexity. High in vitamin C, blackcurrants have long been cherished across central and northern Europe, as well as north Asia, not just for their nutritional value but for their culinary versatility, featured in everything from Crème de Cassis liqueur to preserves, and the classic Kir Royale cocktail. When added to mead, blackcurrants create a beverage that is redolent with the fruit's pleasantly aromatic and slightly sour taste.

Blueberries

In the craft of mead-making, blueberries offer a gentle yet distinctive touch. Sourced from the cooler climes of Michigan, Wisconsin, and Minnesota, these small fruits are a treat for the palate— and a feast for the eyes with their smooth, purple-blue skin. When blueberries meld into the mead, they cast a beautiful purple-pink hue reminiscent of the soft colors at dawn.

Blueberries impart a delicate, fruity flavor that's decidedly berry-like, yet it carries a subtlety not found in the typical blueberry pie. With each pound adding about five fermentation points and an ideal pH balance between 3.11 and 3.22, blueberries infuse the mead with a perfect blend of sweetness tinged with just the right amount of acidity and tannin. Whether you're enjoying them in their prime season (from May to October) or as a year-round delight in mead, blueberries—especially the wild varieties—contribute a strong, vibrant taste.

Cherries

With their rich colors ranging from vibrant reds to golden hues, cherries bring elegance and depth to the traditional mead. These stone fruits, harvested from the orchards of Michigan and Wisconsin and across the varied terrains of Eastern Europe, offer a range of flavors. While sweet cherries are delightful in their own right, the tart varieties genuinely shine in mead-making, infusing the drink with a complexity that sweet cherries can't quite achieve, steering clear of any medicinal undertones.

The influence of cherries in mead extends beyond their flavor, contributing seven fermentation points per pound and boasting a pH range of 3.25 to 4.54. This adds a desirable tartness that balances the mead's sweetness. With their dark, indulgent color, cherries dye the mead a deep reddish-purple, creating a visual allure as captivating as its taste. Perfect for a pie? Yes of course, but in mead, cherries become something more—a fusion of taste and tradition that can elevate a simple beverage into a work of art.

Cranberries

Cranberries, those tart treasures nestled in the acidic bogs of the northern hemisphere, bring a spirited zest to mead crafting. Not just for adornment on a holiday table, these bright red berries, hailing from the historic bogs of New England and the vast expanses of Wisconsin, are celebrated in mead for their sharp, vibrant flavor.

In the careful hands of a mead-maker, cranberries, which yield three fermentation points per pound, can deftly balance the sweetness of honey with their natural acidity, ranging from 2.30 to 2.50 pH. They grow on dwarf shrubs, modest in height but rich in potential, each berry bursting with an intense flavor that infuses the mead with a berry-like aroma and a vibrant red color. The resulting beverage is a visual delight and a taste experience.

Lemons

With its vibrant yellow appeal, the lemon is more than just a staple in the kitchen; it's a transformative ingredient in the world of mead-making. Cultivated in the generous sunshine of Florida and California, this citrus fruit is celebrated for its refreshing zest, the soul of a glass of lemonade! When introduced to mead, the lemon brings not only a touch of its sunny color but also a burst of very crisp tartness, thanks to the citric acid that makes up about 5% of its juice, creating a flavor profile that ranges from the sharpness of a pH of 2.00 to a milder 2.60.

In the delicate dance of mead fermentation, each pound of lemons adds approximately two points, subtly influencing the brew with a sourness that acts as a counterpoint to the inherent sweetness of the honey. Beyond their sour punch, lemons are cherished for their zest (thin, colored outer layer of the skin), which, when used in mead-making, impart essential oils that lend complexity and fragrance, elevating the concoction. The juice, zest, and even the pulp of lemons find their way into the mix, offering mead creators a chance for crafting a drink that's as full of character and refreshing as the fruit.

Mangoes

Mangoes, the jewels of the tropics, infuse mead with a symphony of complex and harmoniously balanced flavours. These succulent fruits, native to Southern Asia, Australia and now flourishing under the golden California sun, offer a delightful sweetness paired with a subtle acidity that can elevate a simple mead to an exotic concoction. The flesh, a vibrant deep orange, is not only juicy and rich in sugar and acid but also exudes a peach/melon-like aroma that can permeate a mead with its heady scent.

Mangoes contribute significantly to the craft of mead-making, adding approximately nine fermentation points (0.009 specific gravity) per pound, and they boast a pH spectrum from 3.40 to 4.80. This allows the mead maker to navigate between tartness and sweetness, achieving a finished product that gleams with an orange-golden hue—a visual promise of the tropical flavors. While mangoes dazzle in various culinary forms, from the creamy depths of ice cream to the refreshing swirl of smoothies, their role in mead is quite transformative, imparting a flavor profile that's as bold as it is intricate, and as inviting as a sun-drenched tropical orchard.

Mulberries

Mulberries bring a distinctive character to the mead-making table with their lush juiciness and resemblance to blackberry drupelets. Deeply rooted in the traditions of Eastern and Central China, these fruits are versatile and flavorful, often savored in their natural state or transformed into rich jellies. The color they lend to mead is a captivating reddish-purple, imbuing the beverage with an appearance that evokes stories of ancient lands and rich soils.

In the delicate craft of fermenting mead, mulberries contribute five fermentation points per pound, and their pH can vary widely from 3.37 to 5.33, depending on the variety. Black mulberries from Southwest Asia and red mulberries from North America are renowned for their robust flavors. In contrast, the East Asian white mulberry is more subtle, offering a gentle taste. This broad spectrum of flavor, from bold to understated, allows the mead-maker to curate a drink with a balance of sweetness and tartness that can occasionally echo the spirited tang of grapefruit.

Oranges

Oranges, the quintessence of citrus, bring more than just a splash of color to the mead-maker's palette; they infuse each batch with the very essence of sunlight! These round, orange-colored fruits, bountifully grown in the orchards of Florida and California, are as versatile in culinary creations as in mead-crafting. Whether enjoyed freshly peeled, zested into vibrant desserts, or pressed into a refreshing juice, oranges are synonymous with freshness.

In the alchemy of mead-making, each pound of oranges contributes six fermentation points, striking a delightful balance between sweet and tart with a pH range from 3.30 to 4.34. The zest is particularly valuable, offering a concentrated burst of essential oils that enrich the mead with complex aromatics, rather than mere sourness. Sweet oranges, savored raw or in juices, and their bitter counterparts, revered in marmalades and preserves, both find their way into meads. They transform the humble honey drink into an elixir that's refreshing, nuanced, and capable of carrying the drinker to the sunny citrus groves in a single sip!

Peaches

With soft, fuzzy skins and hues ranging from a warm blush to sunny gold, peaches are cherished for adding a Southern charm to mead. These round stone fruits originated in China and found a second home in the Southeastern United States. They are a mead-maker's delight, offering a pale yellow flesh that lends a pinkish-golden hue to the drink.

With each pound contributing six fermentation points and a pH that comfortably sits between 3.30 and 4.05, peaches introduce a subtle sweetness to the mead that is distinctly "peachy". When selecting peaches for your mead, consider each variety's unique characteristics—some imbue the mead with a delicate white sweetness. In contrast, others infuse it with a richer golden nectar. Whether enjoyed fresh, baked into pies, or reduced down to a concentrate, peaches bring a versatile, gentle and "summery" fruitiness to mead.

Pears

These fruits, which taper gracefully in a teardrop shape towards the stalk, come in an array of colors from earthy brown to sun-kissed gold, vibrant green, and even a blush of red, with an off-white flesh that's tender and juicy. Originating from the diverse climates of Europe and Asia, pears are versatile in their culinary applications as well as in mead production, favored in everything from fresh, to canned forms to baked goods.

In the process of fermenting mead, pears bring a gentle complexity, offering six fermentation points per pound and a pH ranging from 3.50 to 4.60. They are a modest source of vitamins A and C, potassium and riboflavin, contributing flavor and a touch of nutrition. Whether they're used in their raw state, canned for preservation, or frozen, pears infuse mead with a fruity, understated, and distinctively pear-like essence. This subtlety makes them a preferred choice for a mead that seeks to marry the mellow sweetness of this fruit with the rich depths of honey.

Pineapples

Pineapples, with their distinctive oval to cylindrical shape, bring a burst of tropical flair to the craft of mead-making. Native to the lush regions of South America, Central America, and the Caribbean, this fruit is as vibrant in flavor; beware if its spiky skin! The rind of a pineapple presents a kaleidoscope of colors from dark green to yellow, orange-yellow, or reddish tones when ripe, while the flesh inside ranges from nearly white to yellow, offering a lovely visual display.

Incorporating pineapples into mead introduces quite a sweet, fragrant flavor *and* a subtle tartness coming from the fruit's juicy, fleshy core. With an impressive eight fermentation points per pound and an approximate pH between 3.20 and 4.00, pineapples infuse the mead with a characteristically sharp pineapple essence that is sometimes sweet, sometimes tart—but always refreshing! This melon-like sweetness, combined with the fruit's natural acidity, makes pineapple an exceptional choice for a mead that aims to capture the essence of the tropics!

Raspberries

Raspberries, the jewels of the bramble world, are a cherished addition to the mead-maker's repertoire, bringing a vibrant burst of sweetness and tartness to the honeyed brew. Thriving in the temperate regions stretching from Asia Minor to the forests of North America, these red fruits are traditionally linked to cool climates. They are the stars of jams and tart fillings.

When introduced into mead, raspberries offer more than just a stunning red color that evolves into a deep, inviting shade in the finished drink. Each pound of raspberries contributes around three fermentation points, situating the mead within a pH range that speaks to their innate tartness, between 3.22 and 3.95. It's during the peak of ripeness, when the berries part from the vine with ease and their color is richest, that they are at their sweetest. This ripeness is crucial, as it ensures the mead is endowed with the full spectrum of raspberry flavors—from the berry's natural sweetness to its subtle tannins.

Redcurrants

Redcurrants, those vibrant small berries that dot the landscapes of Northern Europe, offer a splash of color and a wave of tartness to the mead-making process. These berries, known for their bright red appearance, are a staple in creating jellies, pies, and sauces, and they carry that same versatility into the realm of mead.

When redcurrants are introduced into mead, they bring a fermentation contribution of four points per pound and an acidity level ranging from a sharp 2.5 to a more moderate 3.2 pH. This acidity imbues the mead with a crisp, refreshing, and bold tartness. Closely related to their darker relatives, the black ones, redcurrants thrive primarily in Western Europe's cooler climates. Their use in mead is an homage to their culinary heritage, adding a tart flavor profile and a high acidity that can enliven the senses and add a layer of complexity to the brew.

Strawberries

Strawberries carry the essence of spring and early summer to mead-making. Originating from the fields of Europe, these berries have charmed their way into a myriad of culinary delights, from fresh salads to rich jams, to ice creams and decadent pies.

In mead, their contribution is no less significant. The fruit's red hue translates into a subtle pinkish-orange coloration in the final brew, hinting at the gentle flavors infused.

Offering a modest four fermentation points per pound and a pH typically between 3.00 and 3.90, strawberries introduce a sweetness to mead that is more subdued than many other fruits, paired with a soft acidity. This balance is further enriched by tannins from the seeds dotting their exterior—a characteristic unique to the strawberry. With many varieties of this fruit, each brings its subtle twist on the classic strawberry profile.

Other Fruits You Can Use

This stroll through the orchard of melomel-making does not end with the fruits presented in this chapter. They are merely a starting point, a glimpse into the abundant choices at your disposal. Further rich additions await your discovery, for example black raspberry, boysenberry, and the deep notes of elderberry. The tang of key lime and lime, the subtle complexity of marionberry, the tropical burst of passionfruit, the depth of plum, and the desert kiss of prickly pear are all vibrant threads in the melomel tapestry.

Remember, the fruits covered here are popular for a reason—yet they represent only the beginning! The world of melomel is vast and varied, a playground for the adventurous brewer where pretty much any fruit can become the hero of your creation. Let your imagination guide you, and dare to blend, mix, and experiment. With every new batch, there's an opportunity to craft something truly original!

Pyments

Grapes/Credit: mythja (www.Shutterstock.com)

In the realm of pyment creation, the selection of grape varieties is very important. The venerable *Vitis vinifera*, native to the storied wine regions of the Mediterranean and

Central Europe, has been adopted worldwide, a testament to its versatility and broad appeal.

The classification of wines hinges on the composition of grape varieties. Wines predominantly made from a single grape variety, usually 75–85%, are designated varietal wines, a common practice in the United States and Australia. However, blending different varieties is equally esteemed, with many of the most celebrated French wines resulting from such skilled combinations. European wines often carry the name of their region rather than their grape, although specific grape varieties are legally mandated for certain blends.

The significance of terroir cannot be overstated in the production of distinguished wines. Terroir is the sum of all environmental factors, including grape type, vineyard geography, soil quality, climate, and unique local yeast strains, that contribute to a wine's individuality. This concept distinguishes fine wines, even when derived from the same grape. In contrast, mass-produced wines may downplay terroir in favor of consistent, cost-effective profiles.

The subtle flavors and aromas of grape varieties fluctuate with the whims of weather, with descriptors varying from one year to the next. This variability is why different vintages are uniquely valued and sought after. Grape varieties, generally cultivated for either red or white wines, impart their colors and tannins differently—red wines through skin contact and white wines through skin removal at pressing.

In the process of aging, oaking is more frequently employed for reds. This practice contributes character to wines intended for longevity. In assessing pyments, it's essential to distinguish the inherent qualities of the grape juice from those derived from the fermentation process or maturation. Mead-makers may diverge from traditional vintners in their treatment of grapes, prioritizing the varietal's intrinsic characteristics over the established profiles of well-known wines.

Cabernet Sauvignon

Cabernet Sauvignon, renowned for its robust character in the world of wines, is equally esteemed in crafting mead pyments. This red wine grape, flourishing across diverse regions from the sunbathed vineyards of Bordeaux to the fertile valleys of Napa, brings a depth of flavor and color to traditional and dynamic mead. Its versatility shines, as it adapts remarkably well to warm climates, and matures with elegance—especially when aged in oak, a common practice in mead.

The grape colors the mead with a spectrum ranging from dark ruby to a deep, dark purple, almost like bottled twilight. The aroma evokes the countryside, where notes of blackcurrant and dark berries mingle with the rustic scents of leather and tobacco and a subtle undercurrent of cherry. Regarding taste, the Cabernet Sauvignon grape imbues the mead with a full-bodied richness and a tannic backbone that is both assertive and structured. The flavors don't end there; it often carries hints of the very oak that cradles it during aging, adding layers of complexity. Crafting a pyment with

Cabernet Sauvignon is not just an act of fermentation—it's an art that captures the essence of its storied past and global journey.

Chardonnay

Chardonnay grapes, the cornerstone of white viniculture in the United States, also carve an esteemed niche in the crafting of pyment. This grape, which has its roots in the storied soils of Burgundy and has found equal success in the vineyards of California and Australia, is particularly noted for its ability to develop a pronounced varietal character under less-than-ideal growing conditions. Now a byword for premium white wine, Chardonnay commands respect in the world of mead for its refined and nuanced profile.

The subtle characteristics of Chardonnay pair exquisitely with the delicate sweetness of early-season honey. This grape varietal's inherent lightness calls for a nuanced touch; its fragile bouquet of apple, peach, and pear, intermingled with bright citrus and lush tropical fruit notes, can be quickly overshadowed by more dominant flavors.

A discerning mead-maker will appreciate Chardonnay's affinity for adopting oak's warm notes, often evolving into a luscious, buttery complexity with a hint of nuttiness. This grape's adaptability to oak, and its full-bodied yet medium acidity profile leaning towards the drier spectrum, make it a versatile ally in mead-making.

Gewürztraminer

Gewürztraminer grapes, heralding from the esteemed wine-making regions of Alsace and Germany, are a distinguished choice for mead artisans who craft a beverage with depth and a rich sugar profile. The grape's skin, a delicate pink-to-rose hue, imparts a lustrous straw-yellow to golden tint to the mead, infusing it with visual warmth.

The challenge of cultivating Gewürztraminer lies in its early flowering nature and the crucial need for timely harvesting. This precise timing is essential to capture the grape's quintessential flavors and avoid a lacklustre taste from over-ripening. While traditionally leading to sweeter wine finishes, when melded into mead, specific yeast strains can coax the sugars into higher alcohol levels, offering a drier palate experience. In the realm of flavor and aroma, this varietal is nothing short of enchanting, with its bouquet exuding the floral fragrances of rose and honeysuckle, and a flavor profile that touches the tropical notes of lychee, peach, and mango.

Riesling

The Riesling grape, a gem from the cooler winemaking regions like the famed Mosel and the picturesque Alsace, is more than just a fruit—it's a gateway to crafting an exceptional mead. Its subtle yellow-green juice lends a vibrant straw-like clarity to your mead.

Working with Riesling requires a blend of patience and precision. These grapes flourish in the crisp air and thrive on the vine until the moment is right. Harvest too soon, and you miss the sweetness; too late, and the acidity overshadows. But get it right, and you've got a robust yet refined sugar profile—a perfect starting point for fermentation.

Now, let's talk about turning that grape into a really stellar mead. It's a journey from grape to glass that rewards the vigilant! With Riesling, you have the makings of a dry mead, if that's your preference, by choosing yeasts that temper the sweetness into a more subtle alcohol warmth. And the flavors? Imagine a bouquet blooming with the crispness of an apple, the elegance of a rose, and a hint of violet. Each taste is a discovery of fruitiness, where apple meets peach, and apricot dances with pear. With Riesling, you're not just making mead; you're crafting a symphony of taste that can elevate the simplest of gatherings to a memorable affair.

So, as you practise this craft, remember: your meads will be as unique as your methods. Approach each batch with a sense of adventure, a careful eye, and a commitment to quality. Your Riesling mead isn't just a drink; it reflects tradition, skill, and magic when you combine the correct elements. Here's to your mead-making journey—enjoy the first sip of a perfectly balanced Riesling mead!

Other Wine Grapes You Can Use

As we come to the end of this rather fruity part of the book, think of it not as the end—but as a springboard into the vast and varied world of pyment possibilities. The grape varieties we've discussed are just the beginning. Picture the lush, rolling vineyards, each offering a different grape, a different story: the crisp Chenin Blanc, the deep and soulful Merlot, the sun-kissed Muscat, and the earthy Pinot Noir. Imagine the rustic tales that Sangiovese whispers, the bold adventures that Shiraz recounts, and the passionate narratives of Tempranillo.

Then there's Vidal, with its zest for life; Zinfandel, with its heart on its sleeve; the sophisticated charm of Semillon; the radiant Vignoles; and Seyval's refreshing conversation. Each one brings something unique to your mead-making table. But don't stop there. This list is just a conversation starter. Every grape variety holds a promise, an untold potential waiting for your personal touch.

Metheglins

Metheglin—a mead that marries the natural sweetness of honey with the complex aromas of spices—is an age-old concoction that invites both delight and diligence in its creation. Unlike its fruity cousin, the melomel, metheglin does not involve fermentable fruit additives, making it comparatively straightforward to produce. But here's where the craft becomes an art: balancing spices is a delicate endeavor. The potency of spices can quickly overwhelm, leaving no room for error. Every clove, cardamom pod, or cinnamon stick brings an intensity that spoons *and* sensibility must measure. The freshness and quality of your spices are variables that defy the rigidity of recipes.

It's a journey of taste, a test of patience, where a pinch added can mean the difference between a refined sip and an overpowering draft. Remember, you can always introduce more spice, but time or blending with another batch may be your only salvation if you go overboard!

Metheglin/Credit: Kumucia (www.Shutters tock.com)

When it comes to spices, *less* is often more. A heavy-handed metheglin with spices risks being bold and brash, stripping away the mead's nuanced profile. Fruits are far more forgiving; they can be generous in a melomel, imparting a lush, vinous quality or even a liqueur-like richness without overwhelming the senses. Spices, however, demand a disciplined approach. Overdo the ginger or cloves, and you might find your tongue numbed, your palate hijacked—unable to appreciate the subtleties of your next sip. Consider the fiery habanero: revered for its heat, yet in mead-making it must be handled with the utmost respect; a tiny misstep could transform your brew into a culinary conundrum. So proceed with caution, craft carefully, and let your metheglin be a testament to the harmony achieved when spice meets honey in just the right measure.

Sourcing the Finest Spices

Sourcing the right spices for your metheglin is quite like gathering the finest paints for a masterpiece; seeking out vibrant, high-quality ingredients is essential. You won't typically find these premium selections at the local grocery store. Instead, look to specialty spice shops, often found in bustling metropolitan areas or through reputable online vendors. These shops are treasure troves of exceptional spices, and by forging a relationship with the shopkeepers, you'll gain access to the freshest, most potent varieties.

Preparing Your Spices: Preservation and Potency

Preparing your spices is an exercise in timing and technique. Spices, being delicate foliage, have a peak period of aromatic intensity before they begin to lose their luster. To capture their essence, purchase them whole or unground—this applies to allspice, cinnamon, cloves, nutmeg, and more. When you're ready to brew, freshly crack or grind them to unleash their full aromatic potential. This can be done with various tools: a well-designed grinder can save you the hassle of manually processing each spice, while using a mortar and pestle can be a meditative process. However, opting for brass or porcelain is best to avoid retaining scents. Aim for a coarse grind to prevent a cloudy mead; a fine powder can be too easily overdone and lead to sediment.

Adding Spices to Your Metheglin

Tea Bag/Credit: Wild As Light (www.Shutterstock.co m)

As for adding the spices to your brew, think *subtlety and control.* Encase them in a fiber "tea" or a mesh bag, allowing for a manageable flavour infusion. If you're inclined towards thoroughness, consider boiling the bag to sterilize it before it meets your mead. To ensure your spice packet submerges, add a clean marble, stainless steel nut, or bolt—just a little trick to keep everything steeping nicely. The goal is to let the spices mingle gradually with the honeyed liquid, imparting their essence without overwhelming the delicate play of flavors in your metheglin.

When to add the spices

The moment when you decide to introduce spices into your brew isn't "just another step in the process"; it's a pivotal decision that can transform the essence of your mead. Unlike the straightforward path of making a basic mead, metheglin challenges you to think about the impact of timing and technique on your final product.

Traditionally, some have steeped spices during the boil, borrowing from beer-brewing practices. This method allows for an early introduction of flavors but is less common in modern mead-making, where the practice of no-boil techniques prevail. The concern here is not just about losing delicate aromas to the heat but also about hitting that perfect balance of spice without a clear guidepost.

Another approach is to add spices during primary fermentation, immersing them in the young mead when alcohol levels are still low. This method enriches the mead with a deeper spice integration but requires vigilance against potential contamination.

The secondary fermentation stage is a favored time point for many to add spices, leveraging the protective qualities of alcohol to safeguard against bacteria while drawing out the rich flavors of the spices. This method demands patience and a keen sense of taste, as the extraction is gradual and must be monitored to prevent the overshadowing of mead's inherent character.

For those seeking precision and control, making a spice tea or tincture allows immediate flavor assessment and (if necessary) adjustment. These methods, whether boiling water or neutral spirits offer a way to fine-tune the spice level before blending it into the mead. However, they come with their considerations regarding dilution and alcohol content. Amid these varied techniques, the importance of sanitation cannot be overstated, given the potential for contamination in the spice's journey from harvest to your brew. This aspect of mead-making underscores the need for a careful approach, ensuring that the adventures in flavor don't lead to unintended consequences.

Unlocking Spice Flavors in Mead

Unlike the straightforward addition of ingredients in other brews, spices in metheglin require a bit of finesse to unlock their full potential. The medium—water, alcohol, or heat—plays a crucial role in this extraction process. Given that oils aren't a fit for mead's delicate balance, the focus shifts to mastering the interplay between water and alcohol extractions. If you're leaning towards using heat to coax out those intricate flavors, it's a step that needs attention, before fermentation takes the stage.

Picture this: boiling your chosen spices just enough to release their aromatic magic, then weaving this liquid gold into your fermenter. Or perhaps, imagine steeping them in a jar of vodka, letting time work, to then marry this spice-infused spirit later with your mead. Each spice journey adds depth, making your metheglin a drink and a story in a bottle.

Expanding the Metheglin Pantry

This style graciously welcomes a broader spectrum of ingredients beyond the usual herbs and spices. Think of the vibrant hues and textures that flowers, rose petals, or even the boldness of chocolate and coffee can contribute. Venturing further, nuts and chili peppers introduce a blend of warmth and complexity. The essence here is not just in the additions but in understanding how to marry these flavors with the mead's base. It's about recognizing whether these tastes are best drawn out through water or alcohol, and whether they shine brightest in their raw form or with a hint of cooking. Integrating these ingredients thoughtfully, ensuring they can be seamlessly removed after they've imparted their essence, elevates the crafting process.

It's about striking that perfect balance where the added characters blend flawlessly with the mead, enhancing without overpowering, creating a metheglin that sings with the harmony of its components.

Choosing the Right Spices for Your Mead

The canvas expands with every choice as we explore the spices and herbs fit for metheglin. Each addition, from the warmth of cinnamon to the citrusy spark of orange peels, paints your mead with brushstrokes of flavor and aroma. The art lies in selecting spices that resonate with your vision and understanding their influence on the mead's overall character.

Next, we'll explore what makes these selections captivating alongside some thoughtful considerations to remember as you incorporate them into your brew. This stroll through the garden of flavors and aromas promises to enrich your mead-making adventure, ensuring each addition enhances the drink's complexity while maintaining the delicate balance that defines a genuinely *exceptional* mead.

Allspice

When dried, allspice wears a dark reddish-brown cloak and carries the allure of cinnamon, the warmth of nutmeg, the zest of cloves, and a ginger-like snap—all harmonized in a single berry.

As a metheglin-maker, you can harness the heady aromas of allspice to craft a brew that's as complex as it is comforting. Whether your mead leans towards the dry end of the spectrum or you prefer it sweet, allspice fits snugly into your recipe, infusing the mead with a depth of traditional and bold flavour.

The trick is in the balance; this powerful spice should be a gentle murmur, not a shout, in your concoction! A prudent hand will ensure that the allspice complements, not overwhelms, the flavors within your mead. Remember, each berry carries a potency that demands respect—use it wisely, and your metheglin will be the toast of any table!

Chili Peppers

Chili peppers, those vibrant pods from the capsicum family, bring a bold zest to the metheglin repertoire! While they might masquerade as mere fruits, these fiery characters are often cast as spices due to their piquant nature and robust flavors. Imagine—there are over 200 varieties at your fingertips, each with its distinct level of heat and taste, ready to mingle with the sweetness of honey in your mead.

Chilies are like a spirited dance partner to the honey's smooth moves, introducing a lively kick that can elevate a medium-bodied mead to new heights. They are equally adept at cutting through the rich fruitiness of a melomel, adding a layer of complexity that teases the palate.

But the art lies in their measure. The Scoville Scale is the traditional yardstick for gauging their heat, a subjective index that ranges from the mild-mannered to the wildly fiery. This scale isn't an *exact* science, but it's a very helpful guide for navigating the spicy waters and finding your sweet—or spicy—spot.

Patience and precision are your best tools when it's time to bring the heat to your metheglin. Start with a Scoville calculation, adjusting the chili's intensity to the volume of your brew. Given the scale's reliance on human taste buds, this isn't a perfect equation, but it's a solid framework for experimentation. Add chilis incrementally, tasting as you go, and keep meticulous notes. You're seeking that perfect touch of heat that complements but doesn't *overpower*.

And don't forget, when handling these culinary firecrackers, protect yourself! Gloves and thorough cleaning are not just suggestions—they're *essential*. In your adventure with chili peppers, you'll craft a mead that warms the throat and honours the bold spirit of exploration!

Cinnamon

Cinnamon, a spice that evokes memories of comfort and warmth, can transform a metheglin into a drink that feels like a cozy hug. We often buy cinnamon in the States as cassia, a bolder and darker variety. In contrast, true cinnamon, known as Ceylon cinnamon, is lighter, with a delicate sweetness to its flavor. Both come from the tropical trees of the *Lauraceae* family, their barks rolled into quills, with their aromatic oils ready to infuse your metheglin with their essence.

In the craft of metheglin-making, incorporating cinnamon requires a thoughtful touch. For a dry metheglin, where sweetness steps back, cinnamon brings a fullness of flavor to the palate. Matching its intensity with a mead that can stand up to its character is essential, creating a balanced experience. With their soft, anise-like undertones, Delicate Ceylon quills are perfect after just a whisper of spice. When cracked into smaller chips and bundled in a muslin bag, these quills lend themselves to a controlled infusion, allowing you to steep the spice to perfection.

Ginger

This Asian native, a newcomer to the Western spice cabinet, can be found in myriad forms, from its natural rhizome state to dried, candied, or even infused in syrups. Diverse in its uses, ginger is revered in culinary *and* medicinal circles, boasting a robust and astringent flavor.

Fresh ginger, particularly the prized variety from Jamaica, is characterized by its smooth, tan skin and a fresh, zesty aroma that can infuse metheglin with a warm, peppery flavor. While both dried and fresh ginger offer distinct taste experiences, there's a special allure to the fresh form, which imparts a certain vivacity to the mead. When you introduce ginger into your metheglin, its initial boldness gradually mellows, leaving behind a nuanced spice that harmonizes with the natural sweetness of the honey. The amount you add can vary widely, depending on your taste. A conservative hand might opt for a mere ounce or two per five-gallon batch, yielding a subtle hint. In contrast, more adventurous souls might make a bolder approach, scaling up the quantity to satisfy a more pronounced ginger craving!

Black Pepper

Black pepper, the illustrious spice from the vineyards of the East Indies and India's coast, has a rich history, once a commodity rivaling the value of gold. When we speak of peppercorns in metheglin-making, we're not referring to the mild bell pepper found on pizzas but the true *Piper nigrum*. This spice, known for its vibrant pungency and robust aroma, especially when freshly ground, has the unique ability to enhance flavors without masking them, much like its culinary partner—salt.

In the delicate art of metheglin-making, black pepper serves a dual purpose. It can stand as a bold feature or subtly amplify the complexity of other flavors within the mead. Just a gram or two can expand the bouquet of the beverage, allowing different flavors to surface and shine. When pepper is the centerpiece, it imparts an unmistakable earthiness, marrying well with the deeper notes of more aromatic and distinctive kinds of honey like buckwheat or tulip poplar.

However, such a potent spice requires quite a judicious hand. While not every palate yearns for the kick of pepper, for those who appreciate its depth, black pepper can elevate a metheglin from the ordinary to the extraordinary—creating a mead that's not just a drink, but a reflection of ancient trade routes and an opulent culinary history.

Lavender

With its alluring fragrance and delicate purple blooms, lavender brings to metheglin a touch of the Mediterranean's charm. This herb, reminiscent of the gentlest perfume, pairs with honey as naturally as some of life's most iconic duos. When you're looking to capture the essence of lavender in your mead, the secret lies in the timing of the harvest. Aim for that sweet spot right before the blossoms unfurl, or as they begin their display. Choose a day kissed by warmth and the absence of rain, as this is when the blossoms' aromatic oils are most potent.

In the careful craft of metheglin-making, lavender should be added during the secondary fermentation. A modest two to three ounces of these blooms can transform your mead, lending it a subtle, soothing aroma and a hint of color reminiscent of a dawn sky.

Cultivating your lavender can be deeply rewarding for those with green thumbs and a passion for mead. It's worth noting that while lavender thrives in warmth, it can be vulnerable to the chill of northern winters. Protective mulching can help, ensuring that your plants survive to scent your garden year after year. And if you're seeking a hardy variety with a rich scent, "Hidcote" comes highly recommended. Remember, even the leaves have a part to play—if dried, they too can contribute to the layered flavors of your metheglin, making every sip a reflection of your dedication and care.

Vanilla

Vanilla Extract and Beans/Credit: Africa Studio (www .Shutterstock.com)

Vanilla, the cherished essence extracted from the pods of the *Vanilla planifolia* orchid, is not just a flavor—it's an entire experience! This prized spice, cultivated under the nurturing warmth of tropical climates, undergoes a transformative curing process that brings out a rich, sweet aroma and a flavor that's robust, yet devoid of any bitterness.

Introducing vanilla into the delicate process of metheglin-making is like composing a symphony for the senses. The beans are best added in the secondary fermentation, where they quietly work their magic, bolstering the mead's flavor profile with a velvety depth. Vanilla's strength lies in its subtlety—it enhances and rounds out the flavors it accompanies—much like adjusting an equalizer to enrich the sound without stealing the show!

It's particularly effective in fruit-based melomels, where it complements the natural tartness of berries, and in spiced metheglins, where it adds complexity and a certain sophistication. However, vanilla's nuances can be as elusive as they are delightful, blending seamlessly into the mead's complexities. It requires a patient and attentive mead-maker to harness its transient charm, ensuring it contributes to the mead's beautiful range of flavors without fading into obscurity.

Tea

Incorporating tea into metheglin adds a layer of sophistication and depth. Black tea, with its tannic bite and chai, harmoniously blended, each contributes its distinct profile to the mead-making process. Originating from storied lands like India and China, these teas bring a worldliness to the palette of flavors in metheglin—and to your palate!

The tea infusion into your mead should be carefully approached—steeped with intention, *never* added dry. This is due to the natural microbes found on tea leaves, and the less desirable flavors alcohol extraction can yield. A strong brew is critical: one ounce of tea per quart of water, steeped three times as long as usual at a gentle heat, extracts the essence needed for your mead.

It's this richly brewed liquid, devoid of leaves, that you'll want to introduce to your mead, ensuring a smooth integration without an overabundance of tannins.

Deciding *when* to add tea to your mead offers a chance for creativity. Conventional wisdom suggests adding it later in the fermentation to protect those subtle tea notes. To avoid tannins, options like rooibos or chamomile provide a flavorful yet gentle alternative. And for those who subscribe to the belief that timing affects flavor perception—early for an immediate taste, later for a lingering finish—adding tea at various stages might create the layered experience you're after.

This tailored approach, mindful of tradition and innovation, transforms a simple mead into an exquisite metheglin that resonates with every sip.

Chocolate

Chocolate, that luscious product of fermented, roasted, and ground seeds of the tropical cacao tree, can bring a sumptuous richness to metheglin that is both a little indulgent and quite complex. Cocoa nibs (nature's chocolate chips) are ideal for infusing your mead with this beloved flavor. Their introduction to the mead should be timed after the primary fermentation's vigor has waned, ensuring the stirring process doesn't become a bubbly affair.

As the chocolate integrates with the mead, it imparts its dark, soulful essence, though it also leaves behind a trace of cocoa fat. This thin sheen, which may form on the surface, is part of the natural character of chocolate, but is best left out of the final product. Careful racking of the mead beneath this layer ensures that the resulting beverage is infused with chocolate's rich flavor, without the oily residue. This attention to detail in the crafting process ensures that each bottle of chocolate metheglin is a velvety-smooth celebration of one of nature's most exquisite gifts.

Coffee

Let's talk about the heart of your metheglin—the coffee. You might choose to go bold with an espresso or opt for the subtler notes of a cold brew. This choice is pivotal because it's here that the magic happens: the fusion of coffee's robust notes with the yeast's subtleties during fermentation.

If you're leaning towards convenience, instant coffee will be your ally. It's straightforward and fuss-free, although it does speak quietly rather than shout when it comes to flavor. But don't fret—this is where your creativity shines! A sprinkle of spices can elevate the mead, weaving in layers of taste that complement the coffee's murmur.

As for the process, patience is not just a virtue; it's your guide. With a dash of yeast nutrients, you're looking at around three weeks of fermentation. This period is not set in stone; it's more of a dialogue between you and the mead. Is it aromatic enough? Does the taste tell you it's ready? Trust your senses!

And there's more to this than just crafting a drink; it's about versatility. Sweet or dry? Dark or light? Carbonated or still? The choices are endless. And regardless of your path, each bottle of coffee metheglin is a testament to your dedication—a caffeinated chorus paired with honey's subtle, sweet song.

Roses

You'll find that the vast world of roses, from the tens of thousands of varieties that grace gardens from North America to France, offers a spectrum of tastes and aromas to explore.

When adding rose petals to metheglin, you tap into a deep tradition. These petals, ranging in color from the softest whites to the most passionate of reds, are not just a feast for the eyes; they bring with them a delicate, aromatic essence. Each variety, be it the robust Rugosa or the delicately scented tea rose, introduces a distinct character to your brew.

Now, let's get practical. Your cue is after your metheglin's primary fermentation—gently introduce the petals, starting with a suggestion of two ounces. This isn't about overwhelming your creation; it's about enhancing and complementing. If you've frozen the petals beforehand; excellent! You've locked in their peak bloom. And just as you would with any culinary masterpiece, taste and adjust. More petals can always be added, but the key is to proceed with a gentle hand.

Remember, the roses you choose should be as pure as the intentions behind your mead-making! This means being vigilant about avoiding any exposure to pesticides or fungicides. Your roses must be as natural as the honey they'll mingle with.

It's worth doing a little homework, perhaps a chat with a local gardening club or visiting an arboretum, to ensure you get the best—and safest—petals for your potion.

Other Spices

As we reach the end of this spicy, flavorful chapter, it's important to remember that the spices I've shared are merely the beginning of what's possible in the realm of metheglin. Further, there is the wealth of taste within reach: the warming zing of cardamom, the calming whisper of chamomile, the smoky intrigue of chipotle, or the sweet spice of nutmeg, to name but a few. From the fragrant embrace of heather tips to the citrusy zing of lemongrass and the rich, earthy undertones of hazelnut—each spice offers a well-played instrument in the symphony of your mead!

Braggots

Diving into the world of braggot, you'll discover a delightful blend of mead's sweetness with the robust body of malted grains. It's a fascinating beverage that brings out the best of beer and mead, creating a sea of flavors far more intriguing than any single ingredient.

Think of a braggot as a bridge between two rich traditions, a drink honed over centuries through a brewer's craft and resourcefulness. It's the kind of venture that commands respect for the past, and excitement for the personal touch you'll bring in the future.

First Steps in Braggot Making: Extracts or Grains?

If you're getting your feet wet in braggot-brewing, there's a friendly debate to consider: all-grain versus malt extract. All-grain brewing is the artisan's path, giving you control over every nuance from start to finish, but it's a road that requires patience and precision. Malt extracts, on the other hand, are a welcoming shortcut to the world of brewing. They offer a simpler way to achieve the foundation of your braggot without the extensive equipment and time investment.

Especially if you're new to this, malt extracts can be your ally, helping you to build confidence as you learn the ropes. Think of them as your stepping stones to mastering the craft, with each batch a new lesson learned.

Hops in Your Braggot

Hop Pellets Addition into Mead/Credit: Steve Bowers
(www.Shutterstock.com)

And then there's the hoppy horizon! Hops are not mandatory in a braggot, but they are worth your consideration. With their myriad flavors and aromas, hops are like the spice rack of the brewing world; a little here can add a punch of bitterness, a touch there can sprinkle in some floral notes. The trick lies in understanding the timing of their addition and the character of the various hop varieties. But be mindful—hops are sensitive to light. Just as a chef protects a delicate dish from too much heat, shield your hopped braggot from light to keep those "skunky" notes at bay. Each choice you make with hops is a chance to put a signature twist on your braggot, and that's a thrilling prospect.

Step-by-Step Process to Make a Braggot

Roll up your sleeves and get started on your first batch of braggot.

- You'll want to begin with a squeaky-clean workspace and sanitized brewing equipment because cleanliness is paramount in brewing.

- Then, boil a portion of the water you've measured out. Once it's bubbling away, switch off the heat. It's time to introduce the malt extract. Gently stir it in until it's completely dissolved. If you're using hops, this is their cue. Scatter them into the mix, and reheat to a rolling boil. This is a crucial step where the bitterness and aroma of the hops infuse into the liquid, so keep it going for a full hour.

- When your hour is up, you'll have a richly scented wort. Now, let's sweeten the brew. Add the honey, but remember, the heat must be gentle—too hot, and you risk destroying the delicate flavors and aromas. Cool the mixture down, aiming for a temperature that's just right—not too warm, not too cool, but just right for the yeast, somewhere between 65° F and 75° F.

- Pour the cooled wort into your fermenter, which you've already filled with cool water. This is where the transformation begins. Add yeast nutrients and energizer, stirring them to ensure they're well-distributed. Then comes the yeast, previously rehydrated to wake it up from its slumber. Mix it in, and seal your fermenter with a fermentation lock to keep air out and let the gases escape.

- As the yeast gets to work, you'll notice the bubbling of fermentation—it's a sign that everything is going as planned. When the bubbling slows, it's time to transfer your concoction to a secondary fermenter, leaving any sediment behind. Be patient; good things take time, and braggot is no exception!

- Once fermentation is complete, bottle your braggot and give it time to mature.

And remember, the braggot you're brewing is a reflection of your taste and creativity. Feel free to experiment with different flavors. Think about the endless possibilities: cherry, hot pepper, or a coriander-and-orange peel braggot! If the mood strikes, why not give your braggot some sparkle with carbonation? Follow these steps, and you're not just making a beverage—you're crafting a whole experience for you and to share!

CHAPTER 7
Crafting Your Recipe

This chapter is your roadmap to creating a mead recipe that's uniquely yours. Starting out, it's wise to cut your teeth on established recipes—those I will share with you, or ones you've discovered in books, online, from fellow mead enthusiasts, or local homebrew shops. However, as your practice builds and you become more adept in the craft, you'll likely yearn to explore *beyond* these guides.

With experience comes intuition—the ability to gauge just the right amount of each ingredient by feel, making strict adherence to formulas less necessary. However, understanding the foundational steps is crucial.

Here, we look into not just the "how" but also the "why" behind each decision in the recipe development process. You'll learn to navigate the specifics, such as batch size, gravity measurements, and alcohol content, with the same assurance as choosing the perfect honey or special ingredients that make your mead sing.

This chapter is designed to inspire confidence, blending practical advice with encouragement to trust your instincts. As you step into the role of mead composer, remember each batch reflects your individual taste and creativity. Let's launch into mead space with an open heart and a curious mind, ready to create something remarkable!

Step 1: Crafting Your Vision

The mead-making voyage starts with a clear vision. What's the essence of the mead you're dreaming of? Is it the crisp dryness that dances on the palate, the lush sweetness of honey in full bloom, or perhaps a balance in the tender middle? Consider the strength of the brew: do visions of a robust, high-alcohol mead inspire you, or are you inclined towards a gentler, more mellow sip? The choice of honey is pivotal—will it stand alone, or would you fancy a blend of additional flavors? Consider the special touches you'd like to add—whether it's the subtle hint of spices, the boldness of fruits, or the surprises of other ingredients.

This step is about painting a picture of your perfect mead, a blueprint of flavors and sensations that will guide the rest of your brewing adventure.

Step 2: Determining Alcohol Content

Then we stride into the heart of mead crafting—the art of balance and calculation. The alcohol by volume (ABV) is not just a number; it's the spirit of your mead, (pardon the pun!) shaping its character and feel. This magical figure emerges from the dynamic between starting gravity (SG) and final gravity (FG). Remember, as the alchemist of mead-making, you hold sway over only two of these three elements. A simple formula, ABV = 131 * (SG – FG), becomes your wand to magic up the desired strength.

So, starting with an SG of 1.113 and aiming for an FG of 1.005; you're looking at an ABV of around 14.1 percent.

Understand that back-sweetening may alter the gravity readings but won't significantly change the alcohol content, unless it starts a new fermentation round. This step marries the science of brewing with your vision, setting the stage for further creation.

Step 3: Choosing Your Batch Size

Then, the practical aspect of your endeavor comes into play: the batch size. The capacity of your brewing vessels dictates it—aim for harmony between ambition and equipment. The goal is to minimize air exposure during aging, so choosing a batch size slightly larger than your carboy allows for the inevitable losses during racking, ensuring a perfect fill. Whether you're working with standard carboy sizes or contemplating a smaller batch to experiment with, or conserve precious ingredients, this decision frames the scale of your project. It's a balance of practicality and aspiration, guiding how much of your "dream mead" will come to life in this round of brewing!

Step 4: Determining the Amount of Fermentable Sugar and Starting Gravity (SG)

The fourth step in your mead-making involves a crucial decision: determining the amount of fermentable sugar. This decision will significantly influence the body, sweetness, and alcohol content of your final product. The quantity of sugar, primarily sourced from honey in mead-making, is directly tied to the starting gravity (SG) you aim to achieve and the volume of your batch.

To come to grips with this process, you can lean on the starting gravity table below, which serves as a compass guiding you through the landscape of sweetness and strength. This table illustrates how many pounds of honey you'll need per gallon of must (the mead mixture before fermentation) to reach your desired alcohol level. The table correlates these targets with the necessary starting gravity, assuming fermentation reaches a final gravity (FG) of 1.000.

Pounds of Honey per Gallon	Starting Gravity (SG)
0.5	1.02
1	1.04
1.5	1.06
2	1.08
2.5	1.1
3	1.12
3.5	1.14
4	1.16

Step 5: Estimate the Impact of Fruit Addition on Your Mead Gravity

This step addresses an essential aspect of flavoring with fruits. Recognizing that fruit is approximately 85% water by weight, and water weighs on average 8.33 pounds (3.78 kilograms) per gallon is critical to understanding its effect on your mead's starting gravity (SG). Incorporating 9.8 pounds (4.45 kilograms) of fruit essentially adds a gallon (3.79 liters) of water to your batch, diluting the SG more than it contributes to sugars. Knowing this allows for precise adjustments to the recipe to achieve your desired SG, despite the fruit addition.

You will need to calculate the Starting Gravity to target before fruit addition (BFGS – "Before Fruit Starting Gravity").

First, you need to estimate the batch size before fruit addition (BSNF – "Batch Size No-Fruit"):

$$BSNF = TBS - \frac{PWF * FW}{8.33}$$

And the quantity of sugar in fruit (SF).

$$SF = PSF * FW * 46$$

Where:

BSNF = Batch size without fruit addition (Batch size "No-Fruit")

TBS = Total (planned) Batch Size

PWF = Percentage of water in fruit

FW = Fruit weight

PSF = Percentage of sugar in fruit

S F= Sugar from fruit

Then, you can calculate the starting gravity you need to target for your mead before adding fruit.

$$BFSG = \frac{\left(\dfrac{(TBS * (TSG - 1) * 1000) - SF}{BSNF}\right)}{1000} + 1$$

Where:

TSG = Target starting gravity (

SF = Sugar content in fruit

BFSG = Before fruit Starting Gravity

Below is an example:

I want to obtain a mead with a Target Starting Gravity of 1.113, 11 pounds of fruit (4.99 Kg) that is 85% water and 6% sugar, and a total batch size of 5 gallons (19 liters).

TSG = 1.113

FW = 11

PWF = 85%

PSF = 6%

TBS = 5

$$BSNF = 5 - \frac{(85\% \times 11)}{8.33} = 3.877551 \; gallons$$

$$SF = 6\% * 46 * 11 = 30.36$$

$$BFSG = \frac{\left(\dfrac{(5 * (1.113 - 1) * 1000) - 30.36}{3.877551}\right)}{1000} + 1 = 1.138$$

Adjust your initial batch to 3.88 gallons (14.69 liters) with a starting gravity (SG) set at 1.138 to accommodate the addition of fruit. This adjustment ensures that, once the fruit is incorporated, you'll achieve a total must volume of 5 gallons (19 liters) and a SG of 1.113. It's important to remember that the immediate reading on your hydrometer may not reflect this due to the gradual release of water from the fruit. Therefore, prepare your must with only 3.88 gallons at the adjusted SG to offset the fruit's water and sugar content. Consider increasing your batch by a quart or two (approximately a liter or two) to compensate for the volume of fruit pulp that will remain after transferring the mead to the carboy for the secondary fermentation.

Step 6: Determining the Sweetness Finish: Final Gravity

With your starting gravity set, it's time to focus on where your mead will land regarding sweetness—the final gravity (FG). This step is crucial in predicting the sweetness level of your finished mead and understanding how the yeast's alcohol tolerance will interact with the available sugars.

By considering the alcohol by volume (ABV) potential of your chosen yeast strain alongside your target starting gravity (SG), you can forecast the FG your mead will likely achieve. This foresight allows you to anticipate whether the yeast will consume all the sugars before hitting its alcohol tolerance, or stop fermenting due to reaching its alcohol threshold. For instance, if you're aiming for a FG of 1.013 with a yeast that has a 12 per cent alcohol tolerance, the target starting gravity (SG) that you should target is 1.105.

FG = 1.013

ABV = 12%

$$SG = FG + \frac{ABV}{131}$$

$$SG = 1.013 + \frac{12}{131} = 1.105$$

Should the initial gravity fall below the yeast's tolerance threshold, expect the mead to ferment thoroughly, potentially achieving a final gravity of 1.000 or lower, resulting in a dry finish. Conversely, for a sweeter outcome, you can halt fermentation prematurely before the yeast exhausts its potential, or fine-tune the final gravity after fermentation.

Step 7: Choosing the Right Yeast

Your activity in crafting your ideal mead continues with the critical decision of choosing the perfect yeast strain. This step is more than just a formality; it's about understanding the unique contribution each yeast type can make to your brew. Consider each yeast's specific aroma and flavor profiles, nutritional needs, tolerance to varying temperatures, and ability to withstand the alcohol levels you aim for. Also, consider how it performs during fermentation—does it create a lot of foam, or does it settle well, making the brewing process smoother? Making sure the yeast's alcohol tolerance matches your target final gravity is essential for achieving the precise balance and character you desire in your mead.

Step 8: Picking the Perfect Honey

Honey—the very soul of your mead. The choice you make here is profoundly personal and pivotal to the result. Do you want the honey to echo the notes of your added ingredients, or should it stand in bold contrast, adding depth and complexity? Perhaps you're looking for something subtle, a honey that supports without overshadowing. Whether you're drawn to a specific varietal for its standout character or searching for honey with a nuanced, complex profile, the decision will shape your mead's identity. Local honey bring a taste of your region's flora and has the added benefit of being more sustainable and cost-effective, considering the weighty shipping expense. Reflect on the honey profiles discussed earlier to identify the one that best harmonises with your mead's composition, ensuring a blend that resonates with your personal taste and vision.

Step 9: Choosing Your Special Additions

The selection of special ingredients marks a thrilling phase in your mead-making, offering a canvas for your creativity to shine! Moving beyond the foundational elements of honey, water, yeast, and nutrients allows you to infuse your mead with distinct flavors and aromas ranging from the subtle to the bold.

It's important to remember that while the possibilities may seem endless, not *every* ingredient or combination will resonate with *every* palate. Don't feel constrained to simplicity if you're considering adding a single type of fruit for a focused melomel or contemplating a particular spice for a metheglin with a twist. The magic often lies in the synergy of ingredients—how cherries can complement vanilla's warmth, or how peppers' heat can balance raspberries' tartness. From the classic pairing of apples and cinnamon to more adventurous combinations like strawberries and bananas or raspberries and chocolate, the key is finding elements that merge to create a harmonious blend or a delightful contrast, elevating your mead to new heights of complexity and flavor!

CHAPTER 8

Advanced Nutrient Addition Calculations

B eginning the art and craft of mead-making is a bit like setting out on a grand adventure! You start with a simple map—the basic instructions on how to nourish your must with nutrients. It's a solid starting point that works wonders. You'll find that, in this craft, there's a bit of wiggle room for experimentation. And the yeast? Well, they're not too fussy. They'll happily get to work! But here's the thing: there might come a moment when your passion for mead-making takes a new form.

Maybe you're eyeing the commercial horizon, itching to innovate, or driven by a profound dedication to your hobby. That's when you might feel the pull to go deeper, to understand precisely how much nutrition you need, what kind, and when to add it. In the following pages, we will go from following the steps by heart to grasping the *why* behind them. Your mead-making adventure is fast becoming even more rewarding!

Decoding the Lingo

Before we set sail, let's get our bearings by clarifying (or re-clarifying) some terms that will be our companions on this voyage:

- **YAN (Yeast Assimilable Nitrogen)** represents the nitrogen available in the must that yeast cells can utilize to convert sugars into alcohol.

- **Nutrient:** Throughout this discourse, "nutrient" refers exclusively to substances introduced to supplement YAN and micronutrients.

- The essential vitamins and minerals yeast requires for robust growth and successful fermentation are: **micronutrients.**

- **PPM or mg/L (Parts Per Million or Milligrams Per Liter):** These units are used interchangeably to quantify concentrations, illustrating the relationship between a specific nutrient and the overall volume of the must.

PPM—Explained

PPM, or parts per million, is a straightforward measurement indicating the ratio of one component (in our case, YAN) to a million parts of another element (must). To put it in perspective:

1 gram of sugar per liter (1 g/L) translates into 1,000 mg per liter, equivalent to 1,000 PPM.

Understanding Low YAN in Honey

YAN plays a pivotal role in fermentation, acting as the primary nutrient source for yeast. However, honey naturally contains low levels of YAN, typically below 25 parts per million (PPM). This is significantly less than what is considered ideal for fermentation, which can range from 200 to 400 PPM depending on factors like the must's gravity, the yeast strain, and the natural YAN content of other ingredients. Understanding that higher gravity must demand more nitrogen to ensure a smooth and healthy fermentation process is critical to crafting a successful batch of mead.

Risks of Excessive YAN in Mead-Making

The question of whether it's possible to "overdo it" with YAN additions is a valid one. Indeed, while it might seem tempting to generously add nutrients to avoid undernutrition, this approach comes with its challenges. Excessive nitrogen can lead to overly rapid fermentation, characterized by a surge in yeast activity that not only produces undesirable fusel alcohols, known for their harsh "jet fuel" flavor, but also increases the risk of ethyl carbamate formation, a substance recognized as a probable carcinogen. Moreover, commercial mead-makers must navigate legal limits on nutrient additions, making it essential to strike a balance that avoids the pitfalls of under- and over-nutrition.

Consequences of Insufficient YAN

Conversely, the risks associated with *insufficient* YAN cannot be overstated. Skimping on nitrogen can lead to several problematic outcomes, including stuck fermentations, a significant drop in pH, and an uptick in the production of sulfur compounds and fusel alcohols. These issues manifest as off-putting "rotten egg" smells and the notorious "rocket fuel" taste—precisely what every mead-maker aims to avoid!

Understanding the fine line between too much and too little YAN is crucial for ensuring the smooth progress of fermentation, highlighting the importance of careful planning and precision in the art of mead-making.

Calculating YAN Based on Yeast and Sugar Content

Calculating the Yeast Assimilable Nitrogen (YAN) requirement for your mead is an essential step to ensure a successful fermentation process that is tailored to the specific needs of your yeast strain.

Drawing upon the methodology outlined in the *Scott Labs 2016 Winemaking Handbook*, one can determine the necessary nitrogen levels based on the must's sugar content and the yeast strain's nitrogen demands. This approach involves first establishing the sugar concentration in grams per liter (g/L) by utilizing specific gravity and Brix measurements obtained from a hydrometer reading.

Brix, which measures the sucrose content in a solution, indicates that 1° Brix equals 10 grams of sucrose in 1 kilogram of solution, considering the specific gravity to account for the density difference between the sugar-water solution and pure water.

By multiplying the Brix value by the specific gravity and then by 10, you calculate the sugar content in g/L.

This figure is then used with a factor based on the yeast's nitrogen requirement—0.75 for low, 0.90 for medium, and 1.25 for high nitrogen-requiring strains—to calculate the YAN needs in PPM.

Step 1: calculate the sugar content in g/L of your mead recipe

- Sugar g/L = Brix * Specific Gravity * 10

Step 2: Calculate the nitrogen requirement considering the yeast multiplier

- Low nitrogen requiring yeasts: Sugar (g/L) x 0.75

- Medium nitrogen requiring yeasts: Sugar (g/L) x 0.90

- High nitrogen requiring yeasts: Sugar (g/L) x 1.25

If we want to consider the ppm per YAN required, the formulas are:

$$Required\ PPM\ Nitrogens = Brix * Specific\ Gravity * 10 * Yeast\ Multiplier$$

Where the yeast multipliers are:

- Low nitrogen-requiring yeasts: 7.5 ppm YAN per 1°Brix.

- Medium nitrogen-requiring yeasts: 9 ppm YAN per 1°Brix.

- High nitrogen-requiring yeasts: 12.5 ppm YAN per 1°Brix.

Nutrients for Mead-Making: Contributions to YAN

Now that we've determined our YAN requirements, we'll explore the sources from which we can obtain them.

In an earlier section where we delve into the various nutrients suitable for mead making, I highlighted several additives: Go-Ferm, Fermaid O, Fermaid K, and Diammonium Phosphate (DAP). It's best to concentrate on these nutrients for an optimal mead-making process.

Here's a quick rundown of the nitrogen contributions each of these nutrients can provide to your must, measured in grams per liter (g/L) or parts per million (PPM):

- **Diammonium Phosphate (DAP)** enriches your must with a substantial 210 PPM YAN for every gram per liter added, thanks to its 21% nitrogen content.

- **Fermaid K,** with a 10% nitrogen content, offers 100 PPM YAN per gram per liter.

- **Fermaid O** brings 40 PPM YAN to the table for each gram per liter, derived from its 4% nitrogen makeup.

- **Go-Ferm,** containing 3% nitrogen, contributes 30 PPM YAN for every gram per liter added.

This overview will assist you in calculating the precise amounts of each nutrient required to achieve the desired nitrogen levels in your mead must.

Optimizing Fermentation with Staggered Nutrient Additions (SNA)

Wrapping up our discussion, as mentioned before, staggered nutrient additions (SNA) offer a refined solution to overcome the hurdles presented by inorganic nitrogen during fermentation.

Inorganic nitrogen, while readily usable by yeast, can lead to rapid sugar metabolism, causing undesirable temperature surges in the must. Such fluctuations are detrimental to achieving a smooth, clean fermentation, as they can foster the production of fusel alcohols and demand extended aging periods for the mead to mature to a palatable state.

The strategy behind SNA is elegantly straightforward: incrementally introducing nutrients makes it possible to moderate these spikes, thereby ensuring a more consistent fermentation pathway. This method not only minimizes the risk of fusel alcohol formation but also accelerates the timeline for the mead to become pleasantly, wonderfully, drinkable!

SNA typically involves dividing the total nutrient addition into several equal parts, ranging from two to six, and incorporating these portions during the initial stages of fermentation, particularly during aeration phases that span the first one-third to one-half of the fermentation process.

It's important to note that yeast tends to prioritize the consumption of inorganic nitrogen sources when given the choice. Some nutrient schedules recommend transitioning between different nitrogen sources at specific junctures, beginning with a rehydration phase using Go-Ferm, followed by an early addition of organic nitrogen to establish a strong yeast culture, and then shifting to more readily available inorganic nitrogen sources.

The final stages often involve reverting to organic nutrients to sustain yeast activity through the remainder of the fermentation.

Adopting SNA is not a one-size-fits-all solution; it requires customization based on individual needs and the specific characteristics of the mead you are making. Critical to this approach is the understanding that yeast's ability to utilize inorganic nitrogen ceases around an alcohol concentration of 9% ABV, necessitating the early introduction of these nutrients.

This method allows for strategically using nitrogen sources, maximizing yeast health and efficiency while adhering to any desired nutrient limitations. Following a well-defined SNA, as outlined in prior discussions, can significantly enhance the quality and character of the finished mead.

TOSNA: Revolutionizing Mead Nutrient Management

The Tailored Organic Staggered Nutrient Addition, or TOSNA, is an interesting approach to nutrient management in mead-making. This strategy, developed by the insightful team at meadmaderight.com, moves beyond the traditional nutrient addition methods that many mead makers initially learn. What sets TOSNA apart is its adaptability; it's designed to be customized for each specific batch of mead, considering variables such as batch size, initial sugar levels, and yeast variety. This level of customization ensures that each batch of mead can reach its full potential, guided by the expertise of Scott Laboratories, a leading authority in the wine and fermentation industry. The evolution of TOSNA comes from practical applications and refinements in the commercial mead-making sphere—a blend of traditional wisdom and innovative practice.

The Benefits and Mechanics of the TOSNA Protocol

The cornerstone of the TOSNA protocol is the exclusive use of Fermaid-O, prized for its organic nitrogen content. This contrasts sharply with the inorganic nitrogen sources found in other additives like Fermaid-K and DAP (Diammonium Phosphate), which, although effective in accelerating initial fermentation, can often lead to an overly aggressive fermentation process. This vigor, particularly in the early stages, can detract significantly from the mead's flavor and aromatic qualities. By incorporating organic nitrogen through staggered additions, TOSNA ensures a controlled, consistent fermentation process from start to finish.

The claimed benefits of adopting the TOSNA protocol are manifold: it facilitates the achievement of up to 14% alcohol by volume in less than 30 days, promotes a stable and clean fermentation that preserves the mead's delicate aromatic profile, eliminates the risk of fermentation stalling, offers excellent pH buffering, simplifies the management of fermentation temperatures, and reduces the production of sulfur compounds during fermentation. This approach elevates the quality of the final product and enhances the mead-making experience by making the process much more manageable—and predictable!

It's important to highlight that the most recent update to the TOSNA protocol now accommodates using Fermaid K as a viable option.

Step-by-Step Guide

To create a tailored nutrient schedule for your mead using Fermaid-O, you can apply the following calculation:

First, multiply the Brix level by 10, then multiply that result by the nitrogen requirement factor specific to your yeast strain. Divide this total by 50, then multiply by the volume of your batch in gallons to determine the total grams of Fermaid-O needed.

$$Required\ Fermaid\ O\ (grams) = \frac{Brix * 10 * Yeast\ Nitrogen\ Factor}{50 * Batch\ size (gallons)}$$

The Yeast Nitrogen factors are 0.75 for low, 0.90 for medium, and 1.25 for high nitrogen-demanding yeast strains.

This total amount of Fermaid-O should be divided into four equal parts and added to the must at these intervals:

- 24 hours after adding the yeast

- 48 hours after the initial yeast pitch

- 72 hours after the yeast is pitched

- At the point when one-third of the sugar has been fermented, or on Day 7, whichever occurs first.

You'll soon find that precision in nutrient management is critical to crafting those enchanting flavors that define a great mead!

This is where the Tailored Organic Staggered Nutrient Addition (TOSNA) protocol shines, a method I've come to rely on for fine-tuning the nutrients in my recipes. The TOSNA protocol, with its meticulous approach to nutrient calculation, becomes an invaluable tool once you're ready to take your mead-making skills to the next level.

For those eager to explore this method, accessing the TOSNA calculator is straight-forward: scan the QR code below or click the link to https://bit.ly/TOSNA.

What makes the TOSNA calculator particularly user-friendly is its built-in yeast classi-fication system, which categorizes yeasts based on their nutrient requirements. This feature is a game-changer, sparing you the hassle of sifting through yeast supplier websites or flipping back to earlier sections of this book where common mead yeasts are discussed. It's designed to streamline your process, allowing you to focus more on the creative aspects of mead-making while ensuring your brew has everything it needs to really thrive!

So, when you feel confident in your mead-making prowess and are ready to dig deeper into the nuances of recipe formulation, I wholeheartedly recommend trying the TOSNA protocol. It's a decision that could elevate your mead from being very good, to simply unforgettable!

CHAPTER 9
Beginner–Friendly Recipes

W elcome to the final chapter of our exploration into the enchanting world of mead-making!

This chapter is dedicated to you, the aspiring mead-maker, ready to set sail on a voyage of discovery and creation.

Here, I've carefully selected 15 beginner-friendly recipes I've personally tested and wholeheartedly recommend. These recipes are thoughtfully organized into categories we have covered that reflect the diverse styles of mead: from the simplicity of basic meads, rich in tradition and flavor, to the fruit-infused melomels, apple-honey cysers, grape-infused pyments, the beer-mead hybrid braggots, and the spiced metheglins.

Each recipe is designed to be accessible, providing a solid foundation you can build upon as you gain experience and confidence. I encourage you to view these recipes not as final destinations but as starting points.

As you become more familiar with the process, feel empowered to experiment, tweak, and create meads that reflect your taste and creativity.

Let's prepare for this activity, making sure we have all the right tools and knowledge.

- First up, let's talk about batch size. I've designed these recipes with a 5-gallon batch in mind, recognizing that this might not fit everyone's needs or capabilities. If you want to make a smaller batch, say 1 gallon, the math is simple: divide the ingredient quantities by 5. This flexibility allows you to experiment without feeling overwhelmed!

- On to the nutrients. Navigating through nutrient additions can feel like a balancing act. To simplify this, I'm recommending two approaches:

  ○ The first utilizes Fermaid K and DAP, with the "fixed quantities" I recommended earlier in this book (for a standard 5-gallon batch is to use 4 grams of Fermaid K and 8 grams of DAP).

  ○ The second follows the TOSNA method, employing Fermaid O for those looking for an organic alternative. I will indicate the specific Fermaid O and Go-Ferm quantities for each recipe.

- There's no right or wrong choice here; it's really about what resonates with your brewing philosophy. Both paths are tried and true, leading you to mead that will impress.

- Ingredient flexibility is your friend! As you embark on these recipes, remember that substitution is not just allowed; it's encouraged! Whether swapping honey varieties or trying a different yeast, the goal is to make mead that truly *excites* you.

- Don't fret if you can't find the exact ingredients listed—the beauty of mead-making lies in the personal touch you bring to each batch. Embrace the hunt for quality ingredients, and let it be a joyful part of the process rather than a source of stress.

- Understanding the timeline: In the forthcoming recipes, I've chosen not to set strict timelines for each stage of the brewing journey. Instead of watching the clock, I urge you to lean on gravity readings to guide you through the process phases. Let your palate be the final judge, trusting in the taste of your mead to tell you when it's perfectly ready to enjoy.

- Embrace the unpredictability! The path to perfect mead is, of course, paved with trial and error. Many factors influence the final taste, and it's okay if your mead doesn't turn out *exactly* as you envisioned on the first try. Each batch is a learning opportunity, a chance to refine your craft. So, take notes, adjust, and remember that the journey is as rewarding as the destination, that lovely sip!

- Never underestimate the importance of sanitation. Keeping everything clean is the foundation of successful mead-making. It's the one area where precision and diligence pay off, ensuring that your mead is the best it can be.

- While this book serves as your guide, a vast universe of recipes and techniques is out there waiting to be discovered. The recipes included here are merely a snapshot of what's possible. I urge you to venture beyond these pages, to seek out new recipes, to challenge yourself with unique ingredients, and to continue learning and growing as a mead-maker!

Basic Meads

Glasses of Mead/Credit: Aventyr Foto Co (www.Shut terstock.com)

Recipe 1: Basic Dry Mead

Creating a traditional dry mead is all about balance. In this minimalist brew, where yeast works its magic to consume virtually every speck of sugar, the spotlight shines solely on honey, water, and yeast. This simplicity means every step you take, every decision you make, can't be hidden behind residual sweetness or additional flavors.

Crafting a genuinely outstanding dry mead isn't just about mixing ingredients; it's an art form that demands your focus. You'll need to be precise with your recipe, vigilant during fermentation, and scrupulous with sanitation.

Parameters

- Batch Size: 5 gallons
- Starting Gravity (SG): 1.104
- Final Gravity (FG): 1.005
- ABV: 13%

Ingredients

- Yeast: 5 grams Lalvin EC-1118

Nutrients:

- 6.3 grams (0.2 ounces) Go-Ferm

- 18.5 grams (0.7 ounces) of Fermaid-O (TOSNA) or 4 grams (0.14 ounces) of Fermaid K and 8 grams (0.28 ounces) of DAP

- Honey: 14.05 pounds of wildflower honey

- Stabilizers: 5 Campden tablets (sodium metabisulfite) and 2.5 tsp potassium sorbate

Nutrient Addition Schedule

Add nutrients 24, 48, and 72 hours after the initial yeast introduction. The final addition of nutrients should occur either on Day 7 or upon reaching the ⅓ sugar depletion point, depending on which occurs sooner.

Preparation Steps:

- Begin by thoroughly cleaning and sterilizing all equipment you'll be using.

- Preparing the Must:

  - Heat 10 quarts of water in a sizable pot until it reaches a gentle boil.

  - After removing the pot from the stove, gradually mix in the honey, ensuring it fully integrates without any remnants.

  - Introduce extra water to adjust the overall quantity to 5 gallons.

  - Let the must cool down to ambient temperature before taking a gravity measurement.

  - Transfer the must into a fermenting bucket of 7.9 gallons (or bigger), securing it with a lid accommodating an airlock.

- Preparing the Yeast:

  - Prepare the Go-Ferm by combining it with 126 ml of heated water.

  - Wait for the mixture to cool to 104°F, then introduce the yeast, stirring until it dissolves completely.

  - Allow the mix to rest for 15–20 minutes, enabling the yeast to activate and proliferate into a robust colony.

- Primary Fermentation Process:

  - Once the yeast mixture cools to within 10°F of the must's temperature, incorporate it into the primary fermenting bucket, ensuring even distribution.

- Follow the nutrient addition schedule as previously outlined.

- Aerate the must twice daily until the ⅓ sugar break is reached, then reduce to once daily until the ⅔ sugar break.

- Proceed to rack the mead into a secondary vessel once the specific gravity aligns with the targeted final gravity (FG).

- Secondary Fermentation:

  - Transition the must from the primary vessel into a sanitized 5-gallon glass carboy, adding sodium metabisulfite and potassium sorbate to halt yeast activity and clarify the mead by leaving behind most sediment.

  - Seal the carboy with the rubber stopper with an airlock on it

  - Allow the mead to undergo secondary fermentation until it becomes clear.

- Final Racking & Bottling:

  - Rack the mead one more time (in the second carboy) to further clarify until you can read through it. Consider bulk aging before bottling.

Recipe 2: Basic Semi-Sweet Mead

Crafting a semi-sweet mead balances delicately with a final specific gravity between 1.011 and 1.020. Like its dry counterparts, this mead style stands bare, without the cloak of fruits, spices, or additional flavors—except for a whisper of sugar. It's this simplicity that demands your utmost attention to detail.

Making sure your mead is free of off-flavors post-fermentation is not just recommended; it's essential. The absence of other ingredients means there's nowhere for mistakes to hide! Every step must be approached carefully, from selecting ingredients to monitoring fermentation.

Parameters

- Batch Size: 5 gallons

- Starting Gravity (SG): 1.115

- Final Gravity (FG): 1.016

- ABV: 13%

Ingredients

- Yeast: 5 grams Lalvin EC-1118

Nutrients:

- 6.3 grams (0.2 ounces) Go-Ferm

- 20.3 grams (0.8 ounces) of Fermaid-O (TOSNA) or 4 grams (0.14 ounces) of Fermaid K and 8 grams (0.28 ounces) of DAP

- Honey: 15.5 pounds of wildflower honey

- Stabilizers: 5 Campden tablets (sodium metabisulfite) and 2.5 tsp potassium sorbate

Nutrient Addition Schedule & Preparation Steps

For this recipe, you can use the same method and quantities outlined in the previous recipe.

Recipe 3: Basic Sweet Mead

- When we explore sweet meads, we discover a delightful spectrum catering to various palates. This spectrum is segmented into three distinct categories: Sweet, Dessert, and Sack meads.

- Starting with Sweet meads, these concoctions are the gateway to the sweet mead experience. Their final specific gravity falls within the cozy range of 1.021 to 1.030. This level of sweetness brings forth a pleasantly sweet mead without being *overwhelming*, perfect for those who enjoy a subtle hint of nature's nectar.

- Dessert meads take the sweetness up a notch. With a final specific gravity that starts at 1.031 and climbs higher, these meads are the liquid equivalent of a rather decadent dessert. Their richness and depth of flavor make them an ideal finale at the end of a meal, or a treat to savor in moments of indulgence.

- Then we have the Sack meads, renowned for their sweetness and robust Alcohol by Volume (ABV). These meads typically flaunt an ABV ranging from 16–18%, placing them in a league of their own. While they often share the sweetness of a Dessert mead, it's their high alcohol content that truly defines them. This potent combination is crafted with care to ensure a harmonious balance between the alcohol's warmth and sweetness, resulting in a rich, complex, and utterly captivating mead.

Interestingly, the term "Sack" doesn't strictly denote sweetness. A mead can be considered a "Sack" mead if its ABV is 16% or higher, regardless of its sweetness level. However, it's common practice to lean towards sweetness when crafting these high-strength meads. This approach helps in achieving a balanced flavor profile, where the sweetness complements the alcohol, creating a mead that's both nice and potent and very enjoyable!

Parameters

- Batch Size: 5 gallons
- Starting Gravity (SG): 1.124
- Final Gravity (FG): 1.025
- ABV: 13%

Ingredients

- Yeast: 5 grams Lalvin EC-1118

Nutrients:

- 6.3 grams (0.2 ounces) Go-Ferm
- 21.7 grams (0.8 ounces) of Fermaid-O (TOSNA) or 4 grams (0.14 ounces) of Fermaid K and 8 grams (0.28 ounces) of DAP
- Honey: 19 pounds of wildflower honey
- Stabilizers: 5 Campden tablets (sodium metabisulfite) and 2.5 tsp potassium sorbate

Nutrient Addition Schedule & Preparation Steps

You can use the same method and quantities outlined in the previous recipe for this recipe.

Melomels

Cherry Mead/Credit: Nasir hameed siddiqui (www.S hutterstock.com)

In exploring melomel recipes, we will make three crucial gravity measurements to guide us through the process:

- The Initial Starting Gravity—Consider this our starting line. Before introducing any fruit into the mix, this measurement will give us a clear snapshot of where we stand. It's our baseline, the foundation from which our melomel transforms.

- The Predicted Starting Gravity Post-Fruit Addition—This is a bit of a hypothetical scenario. Once our fruits have contributed all their sugar and water to the must, this measurement is where we theoretically land. As we've mentioned, it's important to note that this value is more of an *educated guess.* The nature of fruit means it gradually releases its sugars and water content, making it a challenge to capture this change *accurately* with a hydrometer.

- The Desired Final Gravity—This is our destination, the sweet spot we aim for. The measure tells us when our melomel has reached its perfect balance of sweetness and alcohol content.

Recipe 4: Cherry Melomel

Parameters

- Batch Size: 5 gallons

- Starting Gravity to target before fruit addition: 1.138

- Starting Gravity after fruit addition (non-measurable): 1.107

- Expected Final Gravity (FG): 1.015

227

- ABV: 12%

Ingredients

- Yeast: 10 grams Lalvin 71B

Nutrients:

- 12.5 grams (0.4 ounces) Go-Ferm
- 19 grams (0.7 ounces) of Fermaid-O
- Honey: 13.5 lbs of clover honey (until you read 1.138 on your hydrometer)
- Cherries: 20 lbs of pitted cherries
- 2 tsp acid blend (combination of malic, tartaric, and citric acid)
- 2.5 tsp pectic enzyme
- Stabilizers: 5 Campden tablets (sodium metabisulfite) and 2.5 tsp potassium sorbate

Nutrient Addition Schedule

Add nutrients 24, 48, and 72 hours after the initial yeast introduction. The final addition of nutrients should occur either on Day 7 or upon reaching the ⅓ sugar depletion point, depending on which occurs sooner.

Preparation Steps:

- Begin by thoroughly cleaning and sterilizing all equipment you'll be using.
- Preparing the Must:
  - Heat 10 quarts of water in a sizable pot until it reaches a gentle boil.
  - After removing the pot from the stove, gradually mix in the honey, ensuring it fully integrates without any remnants.
  - Introduce extra water to adjust the overall quantity to 3 gallons (the rest of the water will be added with the cherries)
  - Let the must cool down to ambient temperature before taking a gravity measurement.

- ○ Transfer the must into a fermenting bucket of 7.9 gallons (or bigger), securing it with a lid accommodating an airlock.

- Preparing the Yeast:

  - ○ Prepare the Go-Ferm by combining it with 250ml of heated water.

  - ○ Wait for the mixture to cool to 104°F, then introduce the yeast, stirring until it dissolves completely.

  - ○ Allow the mix to rest for 15–20 minutes, enabling the yeast to activate and proliferate into a robust colony.

- Primary Fermentation Process:

  - ○ Once the yeast mixture cools to within 10°F of the must's temperature, incorporate it into the primary fermenting bucket, ensuring even distribution.

  - ○ Add the cherries (crush them or slice them to release the juice)

  - ○ Add the pectic enzyme

  - ○ Follow the nutrient addition schedule as previously outlined.

  - ○ To tailor the pH to your liking, add Acid Blend.

  - ○ Aerate the must twice daily until the ⅓ sugar break is reached, then reduce to once daily until the ⅔ sugar break.

  - ○ Proceed to rack the mead into a secondary vessel once the specific gravity aligns with the targeted final gravity (FG).

- Secondary Fermentation:

  - ○ Transition the must from the primary vessel into a sanitized 5-gallon glass carboy, adding sodium metabisulfite and potassium sorbate to halt yeast activity and clarify the mead by leaving behind most sediment.

  - ○ Seal the carboy with the rubber stopper with an airlock on it

  - ○ Allow the mead to undergo secondary fermentation until it becomes clear.

- Final Racking & Bottling:

  - ○ Rack the mead one more time (in the second carboy) to further clarify until you can read through it. Consider bulk aging before bottling.

Recipe 5: Blueberry Melomel

Parameters

- Batch Size: 5 gallons
- Starting Gravity to target before fruit addition: 1.122
- Starting Gravity after fruit addition (non-measurable): 1.107
- Expected Final Gravity (FG): 1.015
- ABV: 12%

Ingredients

- Yeast: 10 grams Lalvin 71B

Nutrients:

- 12.5 grams (0.4 ounces) Go-Ferm
- 19 grams (0.7 ounces) of Fermaid-O
- Honey: 15 lbs of clover honey (until you read 1.138 on your hydrometer)
- Blueberries: 10 lbs of blueberries
- 2 tsp acid blend (combination of malic, tartaric, and citric acid)
- 2.5 tsp pectic enzyme
- Stabilizers: 5 Campden tablets (sodium metabisulfite) and 2.5 tsp potassium sorbate

Nutrient Addition Schedule

Add nutrients 24, 48, and 72 hours after the initial yeast introduction. The final addition of nutrients should occur either on Day 7 or upon reaching the ⅓ sugar depletion point, depending on which occurs sooner.

Preparation Steps:

- Begin by thoroughly cleaning and sterilizing all equipment you'll be using.

- Preparing the Must:

  - Heat 10 quarts of water in a sizable pot until it reaches a gentle boil.

  - After removing the pot from the stove, gradually mix in the honey, ensuring it fully integrates without any remnants.

  - Introduce extra water to adjust the overall quantity to 4 gallons (the rest of the water will be added with the blueberries)

  - Let the must cool down to ambient temperature before taking a gravity measurement.

  - Transfer the must into a fermenting bucket of 7.9 gallons (or bigger), securing it with a lid accommodating an airlock.

- Preparing the Yeast:

  - Prepare the Go-Ferm by combining it with 250ml of heated water.

  - Wait for the mixture to cool to 104°F, then introduce the yeast, stirring until it dissolves completely.

  - Allow the mix to rest for 15–20 minutes, enabling the yeast to activate and proliferate into a robust colony.

- Primary Fermentation Process:

  - Once the yeast mixture cools to within 10°F of the must's temperature, incorporate it into the primary fermenting bucket, ensuring even distribution.

  - Add the blueberries (crush them to release the juice)

  - Add the pectic enzyme

  - Follow the nutrient addition schedule as previously outlined.

  - To tailor the pH to your liking, add acid blend.

  - Aerate the must twice daily until the ⅓ sugar break is reached, then reduce to once daily until the ⅔ sugar break.

  - Proceed to rack the mead into a secondary vessel once the specific gravity aligns with the targeted final gravity (FG).

- Secondary Fermentation:

  - Transition the must from the primary vessel into a sanitized 5-gallon glass carboy, adding sodium metabisulfite and potassium sorbate to halt yeast activity and clarify the mead by leaving behind most sediment.

- ○ Seal the carboy with the rubber stopper with an airlock on it

- ○ Allow the mead to undergo secondary fermentation until it becomes clear.

- Final Racking:

  - ○ Rack the mead one more time (in the second carboy) to further clarify until you can read through it. Consider bulk aging before bottling.

Recipe 6: Raspberry Melomel

Parameters

- Batch Size: 5 gallons

- Starting Gravity to target before fruit addition: 1.149

- Starting Gravity after fruit addition (non-measurable): 1.132

- Expected Final Gravity (FG): 1.025

- ABV: 14%

Ingredients

- Yeast: 10 grams Lalvin D47

Nutrients:

- 12.5 grams (0.4 ounces) Go-Ferm

- 23 grams (0.8 ounces) of Fermaid-O

- Honey: 17 lbs of clover honey (until you read 1.149 on your hydrometer)

- Raspberries: 10 lbs of raspberries

- 2 tsp acid blend (combination of malic, tartaric, and citric acid)

- 2.5 tsp pectic enzyme

- Stabilizers: 5 Campden tablets (sodium metabisulfite) and 2.5 tsp potassium sorbate

Nutrient Addition Schedule

Add nutrients at 24, 48, and 72 hours after the initial yeast introduction. The final addition of nutrients should occur either on Day 7 or upon reaching the ⅓ sugar depletion point, depending on which occurs sooner.

Preparation Steps:

- Begin by thoroughly cleaning and sterilizing all equipment you'll be using.
- Preparing the Must:
  - Heat 10 quarts of water in a sizable pot until it reaches a gentle boil.
  - After removing the pot from the stove, gradually mix in the honey, ensuring it fully integrates without any remnants.
  - Introduce extra water to adjust the overall quantity to 4 gallons (the rest of the water will be added with the raspberries).
  - Let the must cool down to ambient temperature before taking a gravity measurement.
  - Transfer the must into a fermenting bucket of 7.9 gallons (or bigger), securing it with a lid accommodating an airlock.
- Preparing the Yeast:
  - Prepare the Go-Ferm by combining it with 250ml of heated water.
  - Wait for the mixture to cool to 104°F, then introduce the yeast, stirring until it dissolves completely.
  - Allow the mix to rest for 15–20 minutes, enabling the yeast to activate and proliferate into a robust colony.
- Primary Fermentation Process:
  - Once the yeast mixture cools to within 10°F of the must's temperature, incorporate it into the primary fermenting bucket, ensuring even distribution.
  - Add the raspberries (crush them to release the juice)
  - Add the pectic enzyme
  - Follow the nutrient addition schedule as previously outlined.

- To tailor the pH to your liking, add acid blend

  - Aerate the must twice daily until the ⅓ sugar break is reached, then reduce to once daily until the ⅔ sugar break.

  - Proceed to rack the mead into a secondary vessel once the specific gravity aligns with the targeted final gravity (FG).

- Secondary Fermentation:

  - Transition the must from the primary vessel into a sanitized 5-gallon glass carboy, adding sodium metabisulfite and potassium sorbate to halt yeast activity and clarify the mead by leaving behind most sediment.

  - Seal the carboy with the rubber stopper with an airlock on it

  - Allow the mead to undergo secondary fermentation until it becomes clear.

- Final Racking:

  - Rack the mead one more time (in the second carboy) to further clarify until you can read through it. Consider bulk aging before bottling.

Recipe 7: Mango Melomel

Parameters

- Batch Size: 5 gallons

- Starting Gravity to target before fruit addition: 1.111

- Starting Gravity after fruit addition (non-measurable): 1.107

- Expected Final Gravity (FG): 1.015

- ABV: 12%

Ingredients

- Yeast: 10 grams Lalvin K1V-1116

Nutrients:

- 12.5 grams (0.4 ounces) Go-Ferm

- 22.8 grams (0.8 ounces) of Fermaid-O

- Honey: 12.5 lbs of clover honey (until you read 1.111 on your hydrometer)

- Mango: 5 lbs of mango slices

- 2 tsp acid blend (combination of malic, tartaric, and citric acid)

- 2.5 tsp pectic enzyme

- Stabilizers: 5 Campden tablets (sodium metabisulfite) and 2.5 tsp potassium sorbate

Nutrient Addition Schedule

Add nutrients 24, 48, and 72 hours after the initial yeast introduction. The final addition of nutrients should occur either on Day 7 or upon reaching the ⅓ sugar depletion point, depending on which occurs sooner.

Preparation Steps:

- Begin by thoroughly cleaning and sterilizing all equipment you'll be using.

- Preparing the Must:

    ○ Heat 10 quarts of water in a sizable pot until it reaches a gentle boil.

    ○ After removing the pot from the stove, gradually mix in the honey, ensuring it fully integrates without any remnants.

    ○ Introduce extra water to adjust the overall quantity to 4.5 gallons (the rest of the water will be added with the mango slices)

    ○ Let the must cool down to ambient temperature before taking a gravity measurement.

    ○ Transfer the must into a fermenting bucket of 7.9 gallons (or bigger), securing it with a lid accommodating an airlock.

- Preparing the Yeast:

    ○ Prepare the Go-Ferm by combining it with 250ml of heated water.

    ○ Wait for the mixture to cool to 104°F, then introduce the yeast, stirring until it dissolves completely.

    ○ Allow the mix to rest for 15–20 minutes, enabling the yeast to activate and proliferate into a robust colony.

- Primary Fermentation Process:

- Once the yeast mixture cools to within 10°F of the must's temperature, incorporate it into the primary fermenting bucket, ensuring even distribution.

- Add the mango slices.

- Add the pectic enzyme.

- Follow the nutrient addition schedule as previously outlined.

- To tailor the pH to your liking, add acid blend.

- Aerate the must twice daily until the ⅓ sugar break is reached, then reduce to once daily until the ⅔ sugar break.

- Proceed to rack the mead into a secondary vessel once the specific gravity aligns with the targeted final gravity (FG).

- Secondary Fermentation:

  - Transition the must from the primary vessel into a sanitized 5-gallon glass carboy, adding sodium metabisulfite and potassium sorbate to halt yeast activity and clarify the mead by leaving behind most sediment.

  - Seal the carboy with the rubber stopper with an airlock on it

  - Allow the mead to undergo secondary fermentation until it becomes clear.

- Final Racking & Bottling:

  - Rack the mead one more time (in the second carboy) to further clarify until you can read through it. Consider bulk aging before bottling.

Recipe 8: Mulberry Melomel (Morat)

Mulberries might not be the first fruit you'd think of, that you can easily pick up at your local grocer and enjoy at your dining table. Their elusive nature and the challenge of picking and processing them often relegates them to a fond childhood memories of adventurous days spent plucking them directly from the tree!

However, these small but mighty fruits hold a secret charm that's especially revealed when they're transformed... into a mulberry melomel. One of the usual hassles of working with these berries—removing the tiny stems—becomes a non-issue.

Parameters

- Batch Size: 5 gallons
- Starting Gravity to target before fruit addition: 1.139
- Starting Gravity after fruit addition (non-measurable): 1.107
- Expected Final Gravity (FG): 1.015
- ABV: 12%

Ingredients

- Yeast: 10 grams Lalvin D47

Nutrients:

- 12.5 grams (0.4 ounces) Go-Ferm
- 19 grams (0.7 ounces) of Fermaid-O
- Honey: 16.5 lbs of clover honey (until you read 1.111 on your hydrometer)
- Mulberries: 15 lbs of mulberries
- 2 tsp acid blend (combination of malic, tartaric, and citric acid)
- 2.5 tsp pectic enzyme
- Stabilizers: 5 Campden tablets (sodium metabisulfite) and 2.5 tsp potassium sorbate

Nutrient Addition Schedule

Add nutrients 24, 48, and 72 hours after the initial yeast introduction. The final addition of nutrients should occur either on Day 7 or upon reaching the ⅓ sugar depletion point, depending on which occurs sooner.

Preparation Steps:

- Begin by thoroughly cleaning and sterilizing all equipment you'll be using.
- Preparing the Must:

- Heat 10 quarts of water in a sizable pot until it reaches a gentle boil.

- After removing the pot from the stove, gradually mix in the honey, ensuring it fully integrates without any remnants.

- Introduce extra water to adjust the overall quantity to 3.5 gallons (the rest of the water will be added with the mulberries)

- Let the must cool down to ambient temperature before taking a gravity measurement.

- Transfer the must into a fermenting bucket of 7.9 gallons (or bigger), securing it with a lid accommodating an airlock.

- Preparing the Yeast:

  - Prepare the Go-Ferm by combining it with 250ml of heated water.

  - Wait for the mixture to cool to 104°F, then introduce the yeast, stirring until it dissolves completely.

  - Allow the mix to rest for 15–20 minutes, enabling the yeast to activate and proliferate into a robust colony.

- Primary Fermentation Process:

  - Once the yeast mixture cools to within 10°F of the must's temperature, incorporate it into the primary fermenting bucket, ensuring even distribution.

  - Add the mulberries (crush them to release the juice)

  - Add the pectic enzyme

  - Follow the nutrient addition schedule as previously outlined.

  - To tailor the pH to your liking, add acid blend.

  - Aerate the must twice daily until the ⅓ sugar break is reached, then reduce to once daily until the ⅔ sugar break.

  - Proceed to rack the mead into a secondary vessel once the specific gravity aligns with the targeted final gravity (FG).

- Secondary Fermentation:

  - Transition the must from the primary vessel into a sanitized 5-gallon glass carboy, adding sodium metabisulfite and potassium sorbate to halt yeast activity and clarify the mead by leaving behind most sediment.

- Seal the carboy with the rubber stopper with an airlock on it

- Allow the mead to undergo secondary fermentation until it becomes clear.

- Final Racking & Bottling:

  - Rack the mead one more time (in the second carboy) to further clarify until you can read through it. Consider bulk aging before bottling.

Recipe 9: Rose Hip Melomel (Rhodomel)

Parameters

- Batch Size: 5 gallons

- Starting Gravity: 1.114

- Expected Final Gravity (FG): 1.015

- ABV: 13%

Ingredients

- Yeast: 10 grams Lalvin 71B

Nutrients:

- 12.5 grams (0.4 ounces) Go-Ferm

- 20.1 grams (0.7 ounces) of Fermaid-O

- Honey: 12.5 lbs of clover honey

- Rose Hips: 20 lbs of dried rose hips

- 2 tsp acid blend (combination of malic, tartaric, and citric acid)

- Stabilizers: 5 Campden tablets (sodium metabisulfite) and 2.5 tsp potassium sorbate

Nutrient Addition Schedule

Add nutrients 24, 48, and 72 hours after the initial yeast introduction. The final addition of nutrients should occur either on Day 7 or upon reaching the ⅓ sugar depletion point, depending on which occurs sooner.

Preparation Steps:

- Begin by thoroughly cleaning and sterilizing all equipment you'll be using.
- Preparing the Must:
  - Heat 10 quarts of water in a sizable pot until it reaches a gentle boil.
  - After removing the pot from the stove, gradually mix in the honey, ensuring it fully integrates without any remnants.
  - Introduce extra water to adjust the overall quantity to 5 gallons
  - Let the must cool down to ambient temperature before taking a gravity measurement.
  - Transfer the must into a fermenting bucket of 7.9 gallons (or bigger), securing it with a lid accommodating an airlock.
- Preparing the Yeast:
  - Prepare the Go-Ferm by combining it with 250ml of heated water.
  - Wait for the mixture to cool to 104°F, then introduce the yeast, stirring until it dissolves completely.
  - Allow the mix to rest for 15–20 minutes, enabling the yeast to activate and proliferate into a robust colony.
- Primary Fermentation Process:
  - Once the yeast mixture cools to within 10°F of the must's temperature, incorporate it into the primary fermenting bucket, ensuring even distribution.
  - Add the dried rose hips.
  - Follow the nutrient addition schedule as previously outlined.
  - To tailor the pH to your liking, add acid blend.
  - Aerate the must twice daily until the ⅓ sugar break is reached, then

reduce to once daily until the ⅔ sugar break.

- ○ Proceed to rack the mead into a secondary vessel once the specific gravity aligns with the targeted final gravity (FG).

- Secondary Fermentation:

  - ○ Transition the must from the primary vessel into a sanitized 5-gallon glass carboy, adding sodium metabisulfite and potassium sorbate to halt yeast activity and clarify the mead by leaving behind most sediment.

  - ○ Seal the carboy with the rubber stopper with an airlock on it

  - ○ Allow the mead to undergo secondary fermentation until it becomes clear.

- Final Racking & Bottling:

  - ○ Rack the mead one more time (in the second carboy) to further clarify until you can read through it. Consider bulk aging before bottling.

Cysers

Apples and Honey/Credit: Kovaleva_Ka (www.Shutte rstock.com)

Crafting an exquisite cyser is like conducting a small orchestra—where the harmonious blend of freshly pressed apple juice, honey, and yeast play together beautifully in time! This delightful concoction is meticulously fermented until it reaches a lovely crisp finish.

Cyser's versatility makes it truly special; it can dance across the entire spectrum, from bone-dry to lusciously sweet.

Recipe 10: Simple Cyser

Parameters

- Batch Size: 5 gallons
- Starting Gravity (SG): 1.122
- Expected Final Gravity (FG): 1.015
- ABV: 13%

Ingredients

- Yeast: 5 grams Lalvin 71B

Nutrients:

- 6.3 grams (0.2 ounces) Go-Ferm
- 21.4 grams (0.8 ounces) of Fermaid-O
- Honey: 10.5 lbs of clover honey (until you read 1.138 on your hydrometer)
- Apple Juice: 4 gallons
- 4 tsp acid blend (combination of malic, tartaric, and citric acid)
- 2.5 tsp pectic enzyme
- Stabilizers: 5 Campden tablets (sodium metabisulfite) and 2.5 tsp potassium sorbate

Nutrient Addition Schedule

Add nutrients at 24, 48, and 72 hours after the initial yeast introduction. The final addition of nutrients should occur either on Day 7 or upon reaching the ⅓ sugar depletion point, depending on which occurs sooner.

Preparation Steps:

- Begin by thoroughly cleaning and sterilizing all equipment you'll be using.

- Preparing the Must:

  - Heat 10 quarts of apple juice in a sizable pot until it reaches a gentle boil.

  - After removing the pot from the stove, gradually mix in the honey, ensuring it fully integrates without any remnants.

  - Introduce extra apple juice to adjust the overall quantity to 5 gallons

  - Let the must cool down to ambient temperature before taking a gravity measurement.

  - Transfer the must into a fermenting bucket of 7.9 gallons (or bigger), securing it with a lid accommodating an airlock.

- Preparing the Yeast:

  - Prepare the Go-Ferm by combining it with 126 ml of heated water.

  - Wait for the mixture to cool to 104°F, then introduce the yeast, stirring until it dissolves completely.

  - Allow the mix to rest for 15–20 minutes, enabling the yeast to activate and proliferate into a robust colony.

- Primary Fermentation Process:

  - Once the yeast mixture cools to within 10°F of the must's temperature, incorporate it into the primary fermenting bucket, ensuring even distribution.

  - Follow the nutrient addition schedule as previously outlined.

  - Add the pectic enzyme

  - To tailor the pH to your liking, add acid blend

  - Aerate the must twice daily until the ⅓ sugar break is reached, then reduce to once daily until the ⅔ sugar break.

  - Proceed to rack the mead into a secondary vessel once the specific gravity aligns with the targeted final gravity (FG).

- Secondary Fermentation:

- Transition the must from the primary vessel into a sanitized 5-gallon glass carboy, adding sodium metabisulfite and potassium sorbate to halt yeast activity and clarify the mead by leaving behind most sediment.

- Seal the carboy with the rubber stopper with an airlock on it

- Allow the mead to undergo secondary fermentation until it becomes clear.

- Final Racking:

  - Rack the mead one more time (in the second carboy) to further clarify until you can read through it. Consider bulk aging before bottling.

Recipe 11: Apple Cinnamon Cyser

Parameters

- Batch Size: 5 gallons

- Starting Gravity (SG): 1.107

- Expected Final Gravity (FG): 1.015

- ABV: 12%

Ingredients

- Yeast: 10 grams Lalvin D47

Nutrients:

- 12.5 grams (0.2 ounces) Go-Ferm

- 19 grams (0.8 ounces) of Fermaid-O

- Honey: 8.5 lbs of clover honey

- Apple Juice: 4 gallons

- 4 tsp acid blend (combination of malic, tartaric, and citric acid)

- 2.5 tsp pectic enzyme

- 5 to 10 cinnamon sticks, to your preference

- Stabilizers: 5 Campden tablets (Sodium Metabisulfite) and 2.5tsp Potassium Sorbate

Nutrient Addition Schedule

Add nutrients at 24, 48, and 72 hours after the initial yeast introduction. The final addition of nutrients should occur either on Day 7 or upon reaching the ⅓ sugar depletion point, depending on which occurs sooner.

Preparation Steps:

- Begin by thoroughly cleaning and sterilizing all equipment you'll be using.
- Preparing the Must:
  - Heat 10 quarts of apple juice in a sizable pot until it reaches a gentle boil.
  - After removing the pot from the stove, gradually mix in the honey, ensuring it fully integrates without any remnants.
  - Introduce extra apple juice to adjust the overall quantity to 5 gallons
  - Let the must cool down to ambient temperature before taking a gravity measurement.
  - Transfer the must into a fermenting bucket of 7.9 gallons (or bigger), securing it with a lid accommodating an airlock.
- Preparing the Yeast:
  - Prepare the Go-Ferm by combining it with 250 ml of heated water.
  - Wait for the mixture to cool to 104°F, then introduce the yeast, stirring until it dissolves completely.
  - Allow the mix to rest for 15–20 minutes, enabling the yeast to activate and proliferate into a robust colony.
- Primary Fermentation Process:
  - Once the yeast mixture cools to within 10°F of the must's temperature, incorporate it into the primary fermenting bucket, ensuring even distribution.
  - Add the cinnamon sticks.
  - To tailor the pH to your liking, add acid blend

- Follow the nutrient addition schedule as previously outlined.

- Aerate the must twice daily until the ⅓ sugar break is reached, then reduce to once daily until the ⅔ sugar break.

- Proceed to rack the mead into a secondary vessel once the specific gravity aligns with the targeted final gravity (FG).

- Secondary Fermentation:

  - Transition the must from the primary vessel into a sanitized 5-gallon glass carboy, adding sodium metabisulfite and potassium sorbate to halt yeast activity and clarify the mead by leaving behind most of the sediment.

  - Seal the carboy with the rubber stopper with an airlock on it

  - Allow the mead to undergo secondary fermentation until it becomes clear.

- Final Racking & Bottling:

  - Rack the mead one more time (in the second carboy) to further clarify until you can read through it. Consider bulk aging before bottling.

Pyments

For those who find themselves equally interested in the rich history of wine *and* the ancient allure of mead, pyments offer a delightful middle ground. These exquisite beverages draw upon honey as their foundational sweetener, weaving it together with one or more varieties of grape blends. This fusion creates a drink that celebrates the *best of both* worlds, marrying the distinct characteristics of wine, and mead, into a singular, harmonious experience!

The beauty of pyments lies in their versatility. They can span the entire sweetness spectrum, catering to every palate, whether you crave something subtly sweet or richly indulgent. Moreover, the choice of grapes introduces an additional layer of complexity, imbuing the mead with a range of finishes from the tannic bite reminiscent of a bold red wine, to the crisp acidity found in white varietals.

Recipe 12: Gewürztraminer Pyment

Parameters

- Batch Size: 5 gallons

- Starting Gravity to target before grapes addition: 1.132

- Starting Gravity after grapes addition (non-measurable): 1.107

- Expected Final Gravity (FG): 1.015

- ABV: 12%

Ingredients

- Yeast: 10 grams Lalvin 71B

Nutrients:

- 12.5 grams (0.4 ounces) Go-Ferm

- 19 grams (0.8 ounces) of Fermaid-O

- Honey: 17.5 lbs of orange blossom honey

- Gewürztraminer Grapes: 27.5 lbs

- 2.5 tsp pectic enzyme/pectinase

- 2 tsp acid blend (combination of malic, tartaric, and citric acid)

- Stabilizers: 5 Campden tablets (sodium metabisulfite) and 2.5 tsp potassium sorbate

Nutrient Addition Schedule

Add nutrients 24, 48, and 72 hours after the initial yeast introduction. The final addition of nutrients should occur either on Day 7 or upon reaching the ⅓ sugar depletion point, depending on which occurs sooner.

Preparation Steps:

- Begin by thoroughly cleaning and sterilizing all equipment you'll be using.

- Preparing the Must:

  ○ Heat 5 quarts of water in a sizable pot until it reaches a gentle boil.

  ○ After removing the pot from the stove, gradually mix in the honey, ensuring it fully integrates without any remnants.

  ○ Crush the grapes and add them to the must

- Add extra water to adjust the overall quantity to 5 gallons

- Let the must cool down to ambient temperature before taking a gravity measurement.

- Transfer the must into a fermenting bucket of 7.9 gallons (or bigger), securing it with a lid accommodating an airlock.

- Preparing the Yeast:

  - Prepare the Go-Ferm by combining it with 250ml of heated water.

  - Wait for the mixture to cool to 104°F, then introduce the yeast, stirring until it dissolves completely.

  - Allow the mix to rest for 15–20 minutes, enabling the yeast to activate and proliferate into a robust colony.

- Primary Fermentation Process:

  - Once the yeast mixture cools to within 10°F of the must's temperature, incorporate it into the primary fermenting bucket, ensuring even distribution.

  - Follow the nutrient addition schedule as previously outlined.

  - Add the pectic enzyme/pectinase

  - To tailor the pH to your liking, add potassium carbonate

  - Aerate the must twice daily until the ⅓ sugar break is reached, then reduce to once daily until the ⅔ sugar break.

  - Proceed to rack the mead into a secondary vessel once the specific gravity aligns with the targeted final gravity (FG).

- Secondary Fermentation:

  - Transition the must from the primary vessel into a sanitized 5-gallon glass carboy, adding sodium metabisulfite and potassium sorbate to halt yeast activity and clarify the mead by leaving behind most of the sediment.

  - Seal the carboy with the rubber stopper with an airlock on it

  - Allow the mead to undergo secondary fermentation until it becomes clear.

- Final Racking:

  - Rack the mead one more time (in the second carboy) to further clarify until you can read through it. Consider bulk aging before bottling.

Braggots

If you're keen on delighting the beer lovers in your circle, venturing into braggot-making could be your next great adventure!

At its core, a braggot marries the essence of honey with the traditional backbone of beer, creating an intriguing and delectable hybrid. This special kind of mead brings together malted barley, grains, and hops alongside the sweetness of honey to forge a drink that captures the hearts of many people exploring the lovely varieties of beer that can be crafted!

Recipe 13: Basic Carbonated Braggot

Parameters

- Batch Size: 5 gallons

Ingredients

- 3 lbs pale liquid malt extract

- 1–2 oz hop pellets

- Yeast: 10 grams of English Ale Yeast

Nutrients:

- 12.5 grams (0.4 ounces) Go-Ferm

- Honey: 4.5 lbs of tupelo blossom honey. You can also consider berry-blossom honey—or simply any honey that you have available.

- 2 tsp citric acid

- 1 priming sugar drop, for bottle carbonation

Preparation Steps:

- Begin by thoroughly cleaning and sterilizing all equipment you'll be using.

- Preparing the Must:

- Start by heating 1.5 quarts of water until it reaches a boil. At the point when the water starts to warm up sufficiently, stir in the pale liquid malt extract to ensure it dissolves completely.

- Add the hops to the mixture and allow it to boil for 55 minutes before incorporating the honey.

- Continue boiling the must for an additional 5 minutes.

- Carefully pour the must directly into a fermentation bucket with a capacity of 7.9 gallons. Add ice-cold water to cool the mixture down until the temperature falls to around 100°F.

- Top off with water at room temperature until you reach a total volume of 5 gallons.

- Add the citric acid.

- Preparing the Yeast:

  - Prepare the Go-Ferm by combining it with 250ml of heated water.

  - Wait for the mixture to cool to 104°F, then introduce the yeast, stirring until it dissolves completely.

  - Allow the mix to rest for 15–20 minutes, enabling the yeast to activate and proliferate into a robust colony.

- Primary Fermentation Process:

  - Once the yeast mixture cools to within 10°F of the must's temperature, incorporate it into the primary fermenting bucket, ensuring even distribution.

  - Cover and let it ferment for a week

- Secondary Fermentation:

  - Transition the must from the primary vessel into a sanitized 5-gallon glass carboy by leaving behind most of the sediment.

  - Seal the carboy with the rubber stopper with an airlock on it.

  - Age in a cool, dark place for at least one to three months.

- Final Racking:

  - Rack the mead one more time (in the second carboy) to further clarify until you can read through it.

- ○ Bottling and Carbonation
- Bottle your mead
  - ○ For carbonation, use one (1) priming drop per 12-ounce bottle. After two days at room temperature in the bottle, let it rest in your basement or another cool spot for 12 days.

Metheglins

Spices Bag/Credit: Weerawat Lomsuk (www.Shutters tock.com)

The world of spices, herbs, and vegetables is an enchanting domain where your mead-making creativity is very much welcomed! This segment of mead-crafting allows you to weave in an array of herbs and spices, turning your brew into a tapestry of taste that's as unique as your own imagination.

Consider this an invitation for your culinary curiosity to lead the way. A whole galaxy of flavors can be made—far beyond the usual suspects. Picture yourself as a culinary explorer, designing a path through uncharted taste territories. The spices and herbs you introduce to your mead could be the first of their kind to sing on someone's palate!

Recipe 14: Ginger Metheglin

Parameters

- Batch Size: 5 gallons
- Starting Gravity: 1.132

- Expected Final Gravity (FG): 1.025
- ABV: 14%

Ingredients

- Yeast: 10 grams Lalvin K1V-1116

Nutrients:

- 12.5 grams (0.4 ounces) Go-Ferm
- 27.5 grams (1 ounce) of Fermaid-O
- Honey: 15.5 lbs of wildflower honey
- 25 ounces of fresh ginger, peeled and sliced thinly. If you also plan to incorporate ginger in the secondary fermentation stage, consider adding another 10 ounces of fresh ginger.
- Stabilizers: 5 Campden tablets (sodium metabisulfite) and 2.5 tsp potassium sorbate

Nutrient Addition Schedule

Add nutrients 24, 48, and 72 hours after the initial yeast introduction. The final addition of nutrients should occur either on Day 7 or upon reaching the ⅓ sugar depletion point, depending on which occurs sooner.

Preparation Steps:

- Begin by thoroughly cleaning and sterilizing all equipment you'll be using.
- Preparing the Must:
  - Heat 10 quarts of water in a sizable pot until it reaches a gentle boil.
  - Add 25 ounces of ginger, ensuring it's peeled and cut into thin slices.
  - After removing the pot from the stove, gradually mix in the honey, ensuring it fully integrates without any remnants.
  - Introduce extra water to adjust the overall quantity to 5 gallons.
  - Let the must cool down to ambient temperature before taking a gravity read.

- Transfer the must into a fermenting bucket of 7.9 gallons (or bigger), securing it with a lid accommodating an airlock. Make sure to filter out the ginger during this step.

- Preparing the Yeast:

  - Prepare the Go-Ferm by combining it with 250 ml of heated water.

  - Wait for the mixture to cool to 104°F, then introduce the yeast, stirring until it dissolves completely.

  - Allow the mix to rest for 15–20 minutes, enabling the yeast to activate and proliferate into a robust colony.

- Primary Fermentation Process:

  - Once the yeast mixture cools to within 10°F of the must's temperature, incorporate it into the primary fermenting bucket, ensuring even distribution.

  - Follow the nutrient addition schedule as previously outlined.

  - Aerate the must twice daily until the ⅓ sugar break is reached, then reduce to once daily until the ⅔ sugar break.

  - Proceed to rack the mead into a secondary vessel once the specific gravity aligns with the targeted final gravity (FG).

- Secondary Fermentation:

  - Transition the must from the primary vessel into a sanitized 5-gallon glass carboy, adding sodium metabisulfite and potassium sorbate to halt yeast activity and clarify the mead by leaving behind most sediment.

  - Add an extra 10 ounces of fresh ginger, peeled and sliced thinly (optional).

  - Seal the carboy with the rubber stopper with an airlock on it.

  - Allow the mead to undergo secondary fermentation until it becomes clear.

- Final Racking & Bottling:

  - Rack the mead one more time (in the second carboy) to further clarify until you can read through it. Consider bulk aging before bottling.

Recipe 15: Vanilla Metheglin

Parameters

- Batch Size: 5 gallons
- Starting Gravity: 1.107
- Expected Final Gravity (FG): 1.015
- ABV: 12%

Ingredients

Yeast: 10 grams Lalvin D47

Nutrients:

- 12.5 grams (0.4 ounces) Go-Ferm
- 19 grams (0.7 ounces) of Fermaid-O
- Honey: 13.5 lbs of orange blossom honey
- 5 tsp of vanilla extract
- 2 ounces of oak cubes (68 cubes)
- Stabilizers: 5 Campden tablets (sodium metabisulfite) and 2.5 tsp potassium sorbate

Nutrient Addition Schedule

Add nutrients 24, 48, and 72 hours after the initial yeast introduction. The final addition of nutrients should occur either on Day 7 or upon reaching the ⅓ sugar depletion point, depending on which occurs sooner.

Preparation Steps:

- Begin by thoroughly cleaning and sterilizing all equipment you'll be using.
- Preparing the Must:
    - Heat 10 quarts of water in a sizable pot until it reaches a gentle boil.

- After removing the pot from the stove, gradually mix in the honey, ensuring it fully integrates without any remnants.

- Introduce extra water to adjust the overall quantity to 5 gallons

- Let the must cool down to ambient temperature before taking a gravity read.

- Transfer the must into a fermenting bucket of 7.9 gallons (or bigger), securing it with a lid accommodating an airlock.

• Preparing the Yeast:

- Prepare the Go-Ferm by combining it with 250 ml of heated water.

- Wait for the mixture to cool to 104°F, then introduce the yeast, stirring until it dissolves completely.

- Allow the mix to rest for 15–20 minutes, enabling the yeast to activate and proliferate into a robust colony.

• Primary Fermentation Process:

- Once the yeast mixture cools to within 10°F of the must's temperature, incorporate it into the primary fermenting bucket, ensuring even distribution.

- Follow the nutrient addition schedule as previously outlined.

- Aerate the must twice daily until the ⅓ sugar break is reached, then reduce to once daily until the ⅔ sugar break.

- Proceed to rack the mead into a secondary vessel once the specific gravity aligns with the targeted final gravity (FG).

• Secondary Fermentation:

- Transition the must from the primary vessel into a sanitized 5-gallon glass carboy, adding sodium metabisulfite and potassium sorbate to halt yeast activity and clarify the mead by leaving behind most sediment.

- Add the vanilla extract.

• Add the oak:

- Disinfect the oak cubes by immersing them in a potassium metabisulfite water solution (a Campden tablet).

- Put the oak cubes in your mead in a sterilized cheesecloth, securing it with a few marbles for weight.

- Seal the carboy with the rubber stopper with an airlock on it

- Allow the mead to undergo secondary fermentation until it becomes clear.

- Final Racking & Bottling:

  - Rack the mead one more time (in the second carboy) to further clarify until you can read through it. Consider bulk aging before bottling.

  - Allow the mead to rest with the oak for approximately 1.5 to 2 months, then start tasting the mead weekly. Transfer the mead into a clean carboy for aging when your preferred oak flavor is achieved.

APPENDIX
Conversion Table

Precision is essential in making mead, and understanding the symphony of measurements is a bit like learning a new language. This appendix offers a comprehensive conversion table to help you with this, ensuring you can effortlessly translate between US customary units and their metric counterparts. This table will be your trusty companion, guiding you through each meticulous step of your brewing adventure.

US Unit	Metric Equivalent
Fluid Conversions	
1 gallon	3.79 liters
33.81 fluid ounce	1 liter
5 gallons	18.93 liters
5.28 gallons	20 liters
1 fluid ounce	29.58 milliliters
1 cup	240 milliliters
0.03 fluid ounces	1 milliliter
1 tablespoon	14.79 milliliters
Mass Conversions	
1 pound	0.45 kilograms
1 ounce	28.35 grams
2.20 pounds	1 kilogram
0.04 ounces	1 gram

For the conversion **from Fahrenheit to Celsius**, you can consider the formula below:

(Degrees Fahrenheit - 32) x (5/9) = Degrees Celsius

Examples:

- Convert 32F into Celsius: (32F – 32F) x (5/9) = 0C

- Convert 104F into Celsius: (104F – 32F) x (5/9) = 40C

CONCLUSION

C ongratulations! Reaching this point in the book, you've either completed your travels through the pages of *Mead-Making for Beginners* or are glancing ahead, eager to see where this path may lead!

Much like beekeeping, mead-making is about nurturing and mastering an art and craft that's both ancient and deeply connected to the natural world. It's a partnership between you and the ingredients you choose to work with, each batch a blend of care, knowledge, and patience.

The rewards of mead-making are twofold. If you take your passion to the next level, there's the tangible delight of tasting your mead and sharing it with friends, family, and possibly even customers. But beyond the colorful bottles and brilliant flavors, there's a deeper satisfaction: you're participating in a tradition that stretches back centuries, contributing to the preservation and innovation of this craft.

Mead has been a centerpiece in gatherings, a muse for poets, and a cherished drink across many societies. By making mead, you're reviving and sustaining these traditions, adding *your* chapter to the ongoing story.

In our 21st century world, where the pace of life often feels relentless, mead-making offers a return to something genuine and grounding. It's a chance to slow down and connect with nature's simplicity and complexity through the alchemy of fermentation.

There's also the joy of experimentation and learning. Whether you're refining your technique, exploring new flavor profiles, or delving into fermentation science, mead-making is a hobby that grows with you.

And so, as you put down this book and perhaps start planning your first (or next) batch of mead, remember that you're not just making a beverage—you're engaging in a practice as old as civilization, offering endless possibilities for discovery and delight.

Keep this book close—not just as a guide, but as a friend on your mead-making voyage! Whether you're an absolute beginner, or becoming more experienced, there's always more to learn, more to try, and more joy in this historic practice.

May your mead-making experience be filled with discovery, enjoyment, and *delicious mead!*

Cheers!

Share Your Mead Mastery: I Value Your Review!

As we draw the curtain on our guide to mead-making performance together, in *Mead Making for Beginners*, I hope you've found this guide to be a treasure trove of knowledge, transforming you from a curious novice into a confident mead maker.

Your thoughts and experiences are precious to me and the entire community of budding mead enthusiasts. By leaving a review on Amazon, you shine the light of guidance for others embarking on their mead-making quest, and also help me fine-tune this guide to serve you and future readers even better in the future!

Scan the QR code below to share your insights. Your support is immensely appreciated, and I look forward to hearing about the remarkable meads you create.

May your brewing journey be filled with exploration, joy, and, most importantly, glasses of really delicious mead!